Advanced Electron Microscopy and Nanomaterials

Advanced Electron Microscopy and Nanomaterials

Selected, peer reviewed papers from the First
Joint Advanced Electron Microscopy School for Nanomaterials
and the Workshop on Nanomaterials (AEM-NANOMAT'09),
Saltillo (Coahuila) México,
September 29th-October 2nd, 2009

Edited by:

Arturo Ponce and Darío Bueno

TRANS TECH PUBLICATIONS LTD
Switzerland • UK • USA

Trans Tech Publications Ltd
Laubisrutistr. 24
CH-8712 Stafa-Zurich
Switzerland
http://www.ttp.net

Volume 644 of
Materials Science Forum
ISSN 0255-5476

Full text available online at *http://www.scientific.net*

***Distributed** worldwide by*

Trans Tech Publications Ltd
Laubisrutistr. 24
CH-8712 Stafa-Zurich
Switzerland

Fax: +41 (44) 922 10 33
e-mail: sales@ttp.net

and in the Americas by

Trans Tech Publications Inc.
PO Box 699, May Street
Enfield, NH 03748
USA

Phone: +1 (603) 632-7377
Fax: +1 (603) 632-5611
e-mail: sales-usa@ttp.net

Saltillo (Coahuila) Mexico
September 29th – October 2nd, 2009

Organizers

Local Committee:

Chairman	Arturo Ponce Pedraza	aponce@ciqa.mx
Co-Chairman	Darío Bueno Baqués	dbueno@ciqa.mx
Program Coordinator:	Leticia Rodríguez González	letyrdz@ciqa.mx
School Secretariat:	Gerardo Martínez and Esmeralda Saucedo	gtadeo@ciqa.mx esaucedo@ciqa.mx
Support Committee:	Oliverio Rodríguez Fernández	oliverio@ciqa.mx
Exhibition contact:	Jesús Olivo Padilla	jolivo@ciqa.mx
Student Support Coordinator:	Luis Ernesto Elizalde Herrera	elizalde@ciqa.mx
Treasurer:	Josefina del Refugio Pérez Huerta	perezh@ciqa.mx

National Committee:

Jesús Arenas Alatorre	Institute of Physics National University of Mexico	jarenas@fisica.unam.mx
Juan Méndez-Nonell	Research Center for Applied Chemistry (CIQA)	jmendez@ciqa.mx
Vicente Garibay Febles	Mexican Petroleum Institute (Mexico)	vgaribay@imp.mx
Leopoldo Enriquez	Jeol de México SA de CV	enriquez@jeol.com.mx
Fernando Mendoza	FEI Company	Fernando.Mendoza@fei.com
Kaheri Arellano	Carl Zeiss de México, SA de CV	k.arellano@zeiss.org

Invited speakers

Dr. Tomas Vystavel FEI Specialist DUAL-BEAM (SEM/FIB) - Netherlands	**Dual/Beam (SEM/FIB) applications** Title 1: TEM Sample Preparation with DualBeam Title 2: Advanced DualBeam Applications in NanoPrototyping and 3-D Characterization
Dr. Dong Tang FEI Specialist HRTEM and STEM - Netherlands	**High resolution transmission electron microscopy (HRTEM) and scanning transmission electron microscopy (STEM)** Title 1: Atomic scale high-resolution transmission electron microscopy Title 2: STEM applications in nano-sciences
M.C. Martin Palacios Jeol Specialist - Mexico	**In-lens and LAB detectors in FEG-STEM** Title 1: The high resolution at low kV in a FEG scanning electron microscope Title 2: Study of nonconductive samples using the r-Filter
Dr. Antonio Casares Carl Zeiss NTS GmbH Specialist EELS and Omega filter - Germany	**Advanced EELS and Omega filter for TEM applications** Title: Advanced EELS and Omega filter for TEM applications
Dr. Marco Antonio Ramirez Carl Zeiss Cross Beam (AURIGA) - México	**Cross Beam (AURIGA)** Title: SEM/FIB preparation samples
Dr. Alfredo Gomez Institute of Physics UNAM Mexico	**HRTEM and electron diffraction simulation** Title: HRTEM and electron diffraction simulation
Dr. Laurence D. Marks Northwestern University - USA	**kinematical and dynamical diffraction and HRTEM** Title: kinematical and dynamical diffraction and HRTEM: quantitative analysis
Dr. Stavros Nicolopoulos NanoMEGAS SPRL - Belgium	**Precession Electron diffraction in TEM** Title: Precession interface Spinning Star
Dr. José Reyes-Gasga Institute of Physics UNAM Mexico	**Electron diffraction** Title: Dynamical electron diffraction in crystals
Dr. Jean-Paul Morniroli Laboratoire de Metallurgie Physique CNRS - France	**LACBED. Applications in Materials Science** Title: LACBED. Applications in Materials Science
Dr. Miguel Ávalos Borja CNyN-UNAM - Mexico	**EELS** Title: EELS theory and applications

International Advisory Committee

Juan Méndez-Nonell	CIQA - (Mexico)
Fernando Ponce	Arizona State University (USA)
Miguel José Yacamán	University of Texas at San Antonio (USA)
János L Lábár	Institute for Technical Physics (Hungary)
Sven Hovmöller	Stockholm University (Sweden)
David Ríos	IPICYT (Mexico)
Robert Sinclair	University of Stanford (USA)
Karla Balzuweit	Universidade Federal de Minas Gerais (Brazil)
Rafael Garía Roja	University of Cadiz (Spain)
José Reyes-Gasga	National University of Mexico (Mexico)
Alejandro Zúñiga	University of Chile (Chile)
Gustavo Martínez	National University of Colombia (Colombia)
Walter Jose Botta	Federal University of São Carlos (Brazil)
Jean-Paul Morniroli	University of Lille (France)
Sergio Lozano-Perez	University of Oxford (UK)

Topics

•Multifunctional nanocomposites
•Smart materials
•Nanoparticles: synthesis and applications
•Structure phenomena and modeling
•Growth of thin films
•Semiconductors and optoelectronic materials
•Other nanomaterials and interdisciplinary topics

Table of Contents

Materials Science Forum Vol. 644 (2010) pp 1-7
© (2010) Trans Tech Publications, Switzerland
doi:10.4028/www.scientific.net/MSF.644.1

Precession Electron Diffraction assisted Orientation Mapping in the Transmission Electron Microscope

Joaquim Portillo[1,3], Edgar F. Rauch [2], Stavros Nicolopoulos[3*], Mauro Gemmi[4] and Daniel Bultreys[3]

[1] SERVEIS Científico-tècnics, Universitat de Barcelona, Catalunya, Spain, qportillo@ub.edu
[2] SIMAP/GPM2 laboratory, CNRS-Grenoble INP, BP 46 101 rue de la Physique, 38402 Saint Martin d'Hères, France
[3] NanoMEGAS SPRL, Boulevard Edmond Machtens No 79, Saint Jean Molenbeek, Brussels, B-1080, Belgium, info@nanomegas.com
[4] Dipartamento di Scienze della Terra A.Desio, Universita degli Studi di Milano, 20133,Milano mauro.gemmi@unimi.it

Keywords: Precession electron diffraction (PED), Electron Backscatter Diffraction (EBSD), Transmission Electron Microscope (TEM) , crystal orientation mapping, phase mapping

Abstract. Precession electron diffraction (PED) is a new promising technique for electron diffraction pattern collection under quasi-kinematical conditions (as in X-ray Diffraction), which enables "ab-initio" solving of crystalline structures of nanocrystals. The PED technique may be used in TEM instruments of voltages 100 to 400 kV and is an effective upgrade of the TEM instrument to a true electron diffractometer. The PED technique, when combined with fast electron diffraction acquisition and pattern matching software techniques, may also be used for the high magnification ultra-fast mapping of variable crystal orientations and phases, similarly to what is achieved with the Electron Backscattered Diffraction (EBSD) technique in Scanning Electron Microscopes (SEM) at lower magnifications and longer acquisition times.

Introduction

Precession electron diffraction, PED, is a new promising technique for electron diffraction pattern collection under quasi-kinematical conditions (as in X-ray Diffraction), which allows solving "ab-initio" crystal structures of nanocrystals. Diffraction patterns are collected while the TEM electron beam is precessing on a cone surface; in this way, only a few reflections are simultaneously excited and, therefore, dynamical effects are strongly reduced. In comparison with normal selected area electron diffraction, PED has the following advantages:

- Oriented ED patterns may be obtained even if the crystal is not perfectly aligned to a particular zone axis. Thus, reliable ED patterns may be collected very quickly without the need to accurately align the observed crystal along a particular zone axis, which is very useful when dealing with beam sensitive materials (zeolites, pharmaceuticals, proteins, etc.)
- Geometry of PED patterns, when including first order Laue zone excitation, show all symmetry elements necessary to determine space and point group of the nanocrystal under investigation, without the need of applying further special techniques, such as Convergent Beam Electron Diffraction.
- PED intensities are quasi-kinematical, likewise to X-rays , up to a thickness of 100 nm and may be used to solve the structure of nanocrystals in a semi-automatic way (Fig.1)
- PED technique may be used in TEM instruments of voltages ranging from 100 to 400 kV and constitutes an effective upgrade of the TEM instrument to a true electron diffractometer (Fig.2).

PED has recently gained large recognition in TEM and X-Ray community and several new nanocrystalline structures have been solved with this technique [1]. On the other hand, a new field of precession diffraction has recently been developed, the orientational/phase mapping through rapid precessed electron diffraction data collection, which may be regarded as the equivalent in the TEM to the Scanning Electron Microscopy based EBSD technique (Electron Backscatter Diffraction)[2]. Effectively, EBSD has gained popularity over the last decade as it provides an effective tool for the characterization of bulk crystalline materials, provided lateral resolution demanded on the sample is not smaller than 500 nm [3].

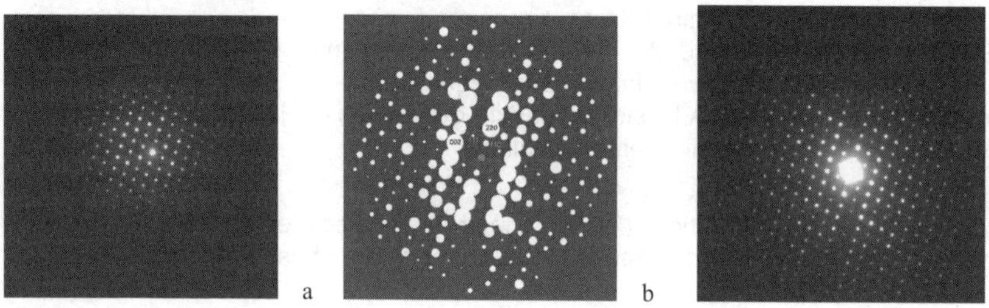

Fig. 1 (a) SAED pattern of spinel MgMoO4 along [1-10], (b) simulated kinematical pattern and (c) quasi-kinematical precession diffraction pattern

If a field emission gun is used in the Scanning Electron Microscope, lateral resolution may be improved down to some 50 nm, and a number of applications have been reported mainly on metals and alloys [4]. The EBSD-SEM technique is based on precise measurements of backscattered Kikuchi line diffraction patterns. However, these lines only appear after incident electron backscattering and double interaction (first inelastic scatter, then elastic diffraction) on surfaces of samples polished to mirror condition, which is a severe limitation on deformed crystalline materials. Moreover, acquisition times, although continously improved, are still on the hour scale.

We have shown that it is possible to apply an analogue of the SEM-EBSD analysis in the TEM, by ultra-fast collection of precessed electron diffraction patterns obtained on a user-selectable area of the TEM with a point spread resolution well within the nanometer range (typically, 10 nm) and without the need of additional surface sample preparation [2]. In this paper we present new advances and results on this technique, PED orientational and phase mapping, with new generation precession instrumentation.

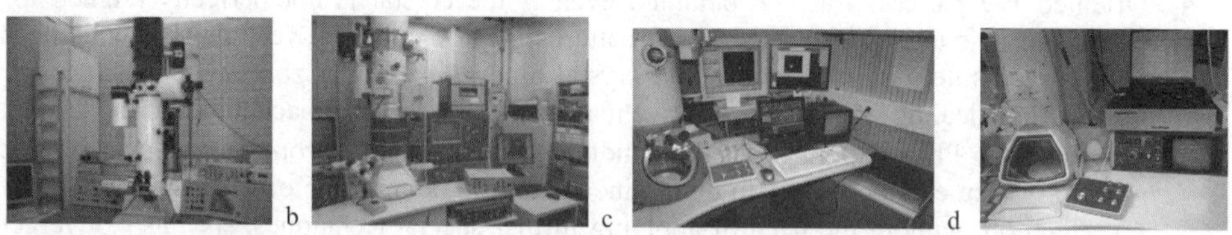

Fig. 2 (a) Precession unit "spinning star" intefaced with Zeiss 912 (120kv), (b) Jeol 2100 FEG (200 Kv); (c) Digital precession unit "DigiSTAR" interfaced with Tecnai 30FEG (300 KV) and (d) with Jeol 2000FX (200 Kv)

New precession instrumentation and methods

A new digital precession instrument has been developed (called DigiSTAR) that can be retrofitted to an old TEM or mounted on a new one, and allows stocking in memory alignment values for driving the beam and image coils simultaneously and at different frequencies and turning (or precession) anglesin the TEM column. It is possible now with this instrument to increase precession angle continously from minimum to maximum with no need to re-align pivot points or readjust descan values, as it is mandatory for the analogue versions of precession instrumentation. Effectively, the most severe limitation imposed on the precession technique is the focused beam deformation and broadening due to several TEM aberrations, namely, spherical and three-fold astigmatism [5]. Thus, the beam spot size diameter is broadened by a factor of at least 3 times when the precession angle increases to 60% of the maximum available on each particular instrument, and at angles of > 2° the spot shape is no longer circular, but highly deformed. The software driving the digital version of the precession instrument enables accurate correction of beam shape and dimension (Fig. 3) even at high precession angles and, therefore, precessed electron diffraction patterns formed on nanocrystals may now be effectively studied under high precession angle illumination for the first time.

Structural information from crystalline materials is present in their electron diffraction patterns, which consist in Kikuchi lines and/or individual diffraction spots. In our Precession Electron Diffraction assisted Orientational and Phase Mapping technique, collected patterns consist exclusively in spots, since Kikuchi lines are smeared out by the high frequency of the precession illumination.

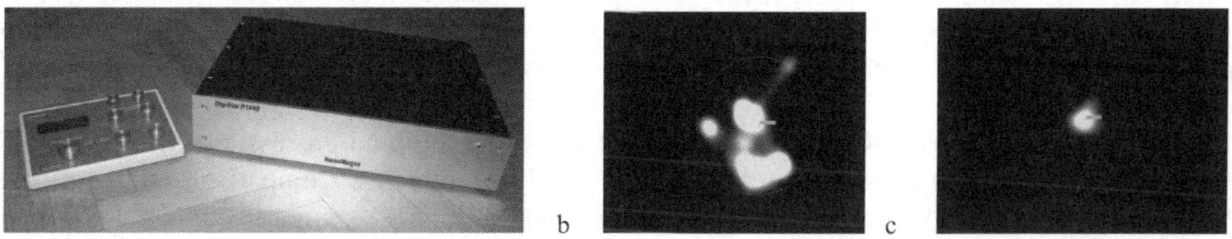

a b c

Fig.3: (a) Digital precession unit "DigiSTAR", shown with manual inteface and control electronics, (b) Deformed beam spot size at precession angle of 2° on a TEM Jeol 2000FX at 8K times magnification, diameter size 1212 nm before software correction, (c) same image after on-line software correction (x8000 magnification, beam diameter 226nm); note beam diameter reduction by factor 6

ED spot patterns are collected sequentially with an ultra-fast optical CCD camera while an area on the sample is simultaneously being scanned by the TEM focused electron beam, which is also being precessed around the direction of incidence at each point. Beam scanning and precessing is controlled by a dedicated external device, which also allows control of beam pivot points and descan pivot points, called "Digistar" and manufactured by the NanoMEGAS company [6]. This device is connected to the beam and image deflector coil control boards present in the TEM. Thus, a TEM retrofitted with this precession device need not necessarily include a scanning option itself. An external ultrafast optical CCD camera, with only 8 bit dynamical range and 250 x 250 pixel image size, mounted in front of the TEM screen is used for the diffraction pattern image collection. This camera records the rapidly changing patterns appearing on the fluorescent screen of the TEM and is the key to a high collection speed since it may work as fast as 180 frames/sec, altough fluorescent screen remanence slightly slows down this performance. During the scanning and precessing of the primary electron beam, thousands of ED spot patterns are recorded and stored in memory of a dedicated computer[2]. In order to proceed with nanocrystal orientation and/or phase identification, each one of the experimental ED spot patterns is compared to one or several sets of thousands of computer generated ED spot patterns, the so-called templates. The software technique for the comparision is based on optimal template matching using cross-correlation (Fig.4).

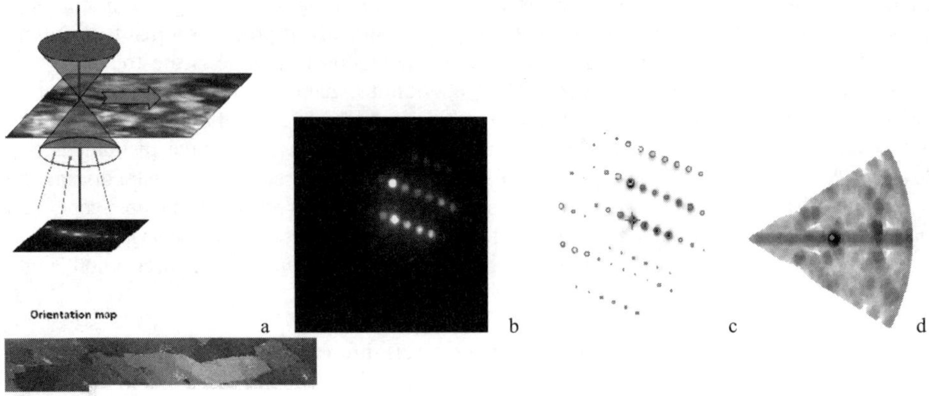

Fig. 4 Schematics of PED assisted Orientational and Phase Mapping: (a) beam scanning with DigiSTAR over a user-defined sample area combined with precession (b) experimental spot PED pattern serial recording in computer memory (c) superposition of individual ED template (red dot pattern) which best matches experimental PED pattern (grey dot pattern), and (d) orientation directional map with grey intensity plot of matching index for the experimental spot PED pattern.

For a typical map of 500 x 300 pixels, the beam scanning (and precessing) over the sample area and associated PED pattern recording may last only 15 minutes. Comparison with simulated templates can be done off-line and takes about 5 minutes for highly symmetric cubic materials and 2 to 24 times longer for unit cells with lower symmetry, because more templates must be generated and compared with experimental ED spot patterns for the same angular resolution (typically 1°). The scanning step chosen for the rastering is about half the size of the primary electron beam size, that is 12 nm for a 25 nm spot size on a TEM with LaB_6 electron gun, and the resulting lateral resolution of the obtained map will be of the order of the latter value. Template generation is done on the basis of unit cell dimensions and geometry as well as inherent atomic positions, for each of the known phases present in the examined sample. Using this data a complete spatial set of ED spot patterns is generated under purely kinematical conditions. The comparison of these templates with experimental ED spot patterns is run searching a maximum match for spot positions and their relative intensities, and the parameter quantifying the match is called the correlation index.

The software developed specifically for the technique allows obtention of crystal orientation maps, virtual bright field maps, correlation index maps and reliability index maps. The latter are related to the statistical significance of the orientational assignation chosen for each point of the map.

Essentially, the virtual brigth field map is obtained by plotting the intensity fluctuations of the central spot in the PED spattern. Such maps are often more helpful for comparision with the final orientation map of the scanned area than real bright-field TEM images of the same area, since these are usually suffering from diffraction contrast and curvature contrast (which fades significantly under precessed illumination). Correlation index maps are mainly used to emphasize structural details such as crystals having different orientations. The reliability index, which is analogous to the SEM-EBSD confidence index, is defined in such a way that it has a minimum when more than one solution is proposed. Such reliability maps clearly reveal grain boundaries and textures (Fig. 5).

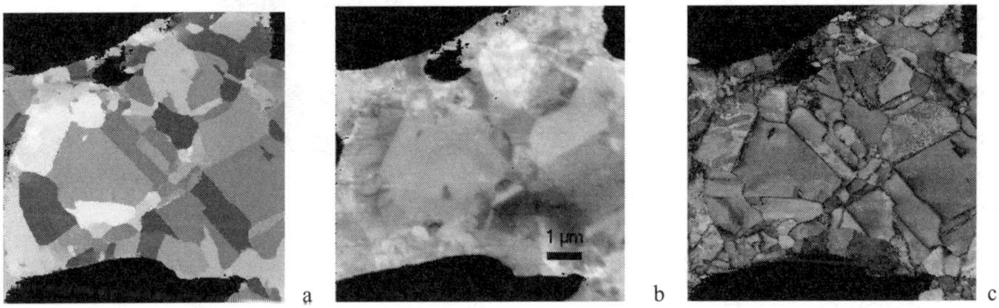

Fig. 5 (a) orientation map revealing nanotwins in Cu thin foil sample (Jeol 3010 LaB6 TEM, 300KV , spot size 12 nm, step 5nm (b) pseudo-bright field area of the same sample (c) Reability map showing clearly grain boundaries.

Although orientation and phase mapping coupled with template matching works satisfactorily for a number of materials without using precession of the primary electron beam [6,7], some limitations are present in the form of ambigous orientation/phase maps due to the poor quality of experimentally acquired ED spot patterns. This is the case for thick crystals where ED patterns show a combination of a strong diffuse inelastic scattering background with Kikuchi lines and a reduced number of faint diffraction spots. For such thick crystals and in general, precessed ED patterns exhibit a higher number of ED spots with especific intensity distribution and do not show Kikuchi lines. Consequently, the template matching software routine will produce higher values of the correlation index when using PED patterns, and orientation maps will exhibit less ambiguity.

A specific experiment was designed in order to outline the influence of beam precession on template matching/indexing for local orientation determination. A series of ED patterns were acquired on a single nanocrystal of mayenite, with formula $Ca_{12}Al_{14}O_{33}$ and cubic unit cell of 11.94 Å, starting from an initial electron beam incidence parallel to the (100) direction in the crystal and then tilting the TEM sample holder along its axis off this (100) direction in steps of 0.05° until a final off-initial incidence direction of 18° was attained. The series of experimental ED patterns were recorded twice, first without and then with a precession angle of 0.35°, while keeping the beam stationary, unrastered on the same point of the crystal. Over 400 diffraction patterns were collected for each tilt series and then analysed with the template matching software. The results are shown as misorientations calculated for every experimental pattern with respect to the initial orientation. The origin of both graphs starts in 0° misorientation for the first data point and should extend towards 18° for the last measured data point (number 400). The blue unprecessed data show significant ambigous orientation assignation, while the red precessed data nicely scale along a constant slope. The stepwise shape of the red curve is due to the fact that the database is generated in orientation steps of half a degree. When comparing the experimental spot ED patterns shown in figure 6, it is worth remarking the absence of Kikuchi lines and the increased number of spots present in the precessed pattern compared to the unprecessed pattern, which explains the cause of the orientation assignation ambiguity in the blue curve.

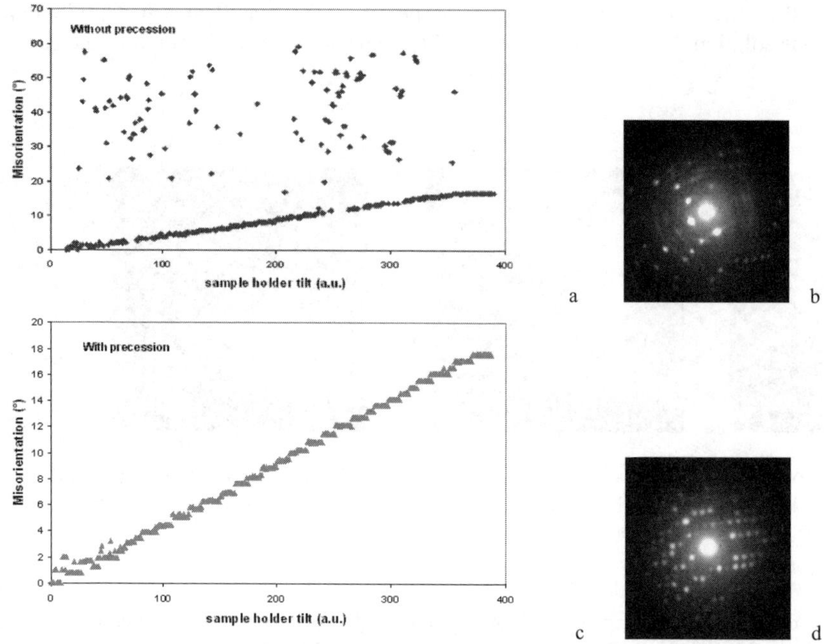

Fig. 6 (a) Orientation values of mayenite $Ca_{12}Al_{14}O_{33}$ (100) nanocrystal obtained after tilting it off this axis from 0 to18° with step of 0.05° along TEM sample holder direction; orientation values obtained by pattern matching-technique applied over set of simulated templates containing whole angular range (b) experimental spot ED pattern originating one point in blue curve, (c) the same experiment with 0.25° precession angle, (d) experimental spot ED pattern originating one point in red curve (note absence of Kikuchi lines and increased number of spots)

This demonstrates that even at small precession small angles, a substantial quality improvement is obtained in orientation maps obtained in the TEM when activating precession illumination.

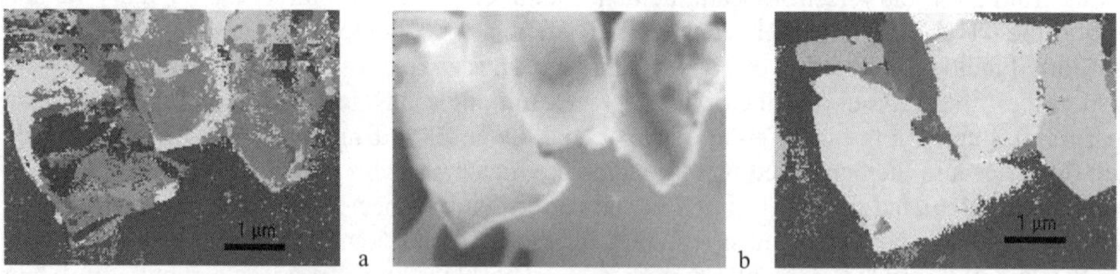

FIG.7 (a) Orientation mapping of mayenite crystals without precession (b) virtual bright field and (c) orientation map obtained with precession angle of 0.35°

This clear improvement of orientation determination can also be shown in the maps of Fig. 7 where an area that contains randomly oriented mayenite crystals was scanned twice (with and without precession). The orientation map in Fig. 7a shows color fluctuations indicating frequent misassigned patterns because of their low quality when compared with simulated templates. The map in Fig. 7c was obtained under the same conditions as in Fig.7a but using precession at 0.35° angle during beam scanning. As it can be observed, orientation map quality increases with precession, since each colour domain corresponds to a single crystal grain with a definite orientation and should not show colour fluctuations, as it is the case.

Conclusions

Precession Electron Diffraction assisted Orientation Mapping is a novel automated crystal phase and orientation mapping technique that benefits from the high spatial resolution (nm size) of TEM. This novel technique has been implemented onto a single instrument called ASTAR which enables beam precession along incident direction and simultaneous scanning over a micrometer or nanometer sized sample area. It can successfully be mounted onto new or old TEMs and uses beam size and shape correction via a digital precession device, termed DigiSTAR.

Acknowledgements

Samples for Fig. 5 were supplied by Prof. N. Llorca of the University of Barcelona and her collaboration on this project is acknowledged.

References

[1] Ultramicroscopy, special issue: New Frontiers in electron crystallography June/July 2007, vol.107,issues 6-7

[2] E.F. Rauch, M.Veron, J.Portillo, D.Bultreys, Y.Maniette and S.Nicolopoulos, "Automatic crystal orientation and phase mapping in TEM by precession diffraction", Microsc. and Anal. **93** (2008), S5-S8

[3] V.Randle, O.Engler, "Introduction to texture Analysis: Macrotexture, Microtexture and Orientation Mapping", CRC Press, 2000

[4] D.P.Field, S.I.Wright, D.J.Dingley, Proc. 11th International Conference on Textures of Materials (ICOTOM11), Ed. Z. Liang et al, International Academic Publishers, China, 94

[5] C. Own, PhD dissertation, Northwestern University, Chicago, December 2008, http://www.numis.northwestern.edu/Research/Current/precession

[6] E.F. Rauch, L. Dupuy, " Rapid spot diffraction pattern identification through template matching", Arch. Metall. Mater **50** (2005), 87-99.

[7] E.F. Rauch, M. Veron, "Coupled microstructural observations of local texture measurements with an automated crystallographic orientation mapping tool attached to a TEM", J.Mater. Sci. Eng. Tech. **36** (2005) 552-556.

[8] www.nanomegas.com, NanoMEGAS-CNRS Patent pending technique.

Materials Science Forum Vol. 644 (2010) pp 9-12
© (2010) Trans Tech Publications, Switzerland
doi:10.4028/www.scientific.net/MSF.644.9

TEM characterization on the nanocomposite Al 7075 and silver nanoparticles synthesized by powder metallurgy

R. Flores-Campos[1,2,a], R. Deaquino-Lara[1,3,b], J.M. Herrera-Ramírez[1,c], A. Ponce[4,d], I. Estrada-Guel[1,e], M. Miki-Yoshida[1,f], R. Martínez-Sánchez[1,g*].

[1] *Centro de Investigación en Materiales Avanzados (CIMAV), Laboratorio Nacional de Nanotecnología, Miguel de Cervantes No. 120, CP 31109, Chihuahua, Chih., México.*
[2] *Departamento de Ingeniería. Tecnológico de Monterrey Campus Saltillo, Prol. Juan de la Barrera No. 1241 Ote., Col. Cumbres, CP 25270, Saltillo, Coah., México.*
[3] *Centro de Investigación y de Estudios Avanzados Unidad Saltillo, Carretera Saltillo-Monterrey Km 13, Col. Molinos del Rey, CP 25900, Ramos Arizpe, Coah., México.*
[4] *Centro de Investigación en Química Aplicada, Blvd. Enrique Reyna Hermosillo No. 140, CP 25253, Saltillo, Coahuila., México.*
[a]ruben.flores@itesm.mx, [b]rogelio.deaquino@cinvestav.edu.mx, [c]martin.herrera@cimav.edu.mx, [d]aponce@ciqa.mx, [e]ivanovich.estrada@cimav.edu.mx, [f]mario.miki@cimav.edu.mx, [g]roberto.martinez@cimav.edu.mx

Keywords: nanocomposites, mechanical milling, aluminum alloys.

Abstract. Aluminum-based nanocomposites have been produced by mechanical milling, introducing silver nanoparticles within the matrix of a 7075 aluminum alloy using a high energy ball mill. The milled products were compacted by uniaxial load and pressure-less sintered under argon atmosphere, and finally hot extruded. Silver nanoparticles are well dispersed into the matrix of the powder particles as well as in the matrix of the extruded material. Transmission electron microscopy (TEM) analyses are used to corroborate and understand the hypothesis that second-phase particles finely and homogeneously dispersed in the matrix give greater strength to the material. In addition to the strengthening effect, the nanoparticles act like a process control agent (PCA) since the crystallite size of the nanocomposite is smaller at higher contents of nanoparticles.

Introduction

Due to its low density and good workability aluminum alloys have a great range of industrial applications [1]. The special mechanical properties of 7XXX series aluminum alloys make them suitable for special applications in aerospace industries. Nevertheless, these properties demanded by the aerospace industry are every day of higher magnitude. Composite materials open the possibility to cover these industrial needs. The reinforcement of an aluminum alloy matrix with second-phase particles produces a strong composite that combine the properties of the matrix and the reinforcement particle [2-6]. Composite materials which involve nanostructures have been called nanocomposites. Mechanical milling (MM) and mechanical alloying (MA) are routes to produce nanocomposites with good physical and mechanical properties [7]. The characterization of these materials is done with different methods; the structural properties are evaluated with the use of TEM, one of the most important techniques to study nanomaterials [8], among others.
The focus in this work was the structural and mechanical characterization of Al_{7075} nanocomposites produced by introducing carbon coated silver nanoparticles into the matrix of the aluminum alloy, by milling process. The mechanical properties were obtained by tensile tests and the yield stress (σ_y) and the ultimate tensile strength (UTS) were determined. The microstructural study was done in both, as-milled and as-extruded conditions by TEM and X-ray diffraction (XRD).

Experimental procedure

7075 aluminum (Al_{7075}) in a metal shaving condition and carbon coated silver nanoparticles (Ag-C NP) were the starting materials. The Ag-C NP is a product obtained from Nanotechnologies, Inc. (Austin, TX), which is produced by arc discharge and stabilized with carbon from a hydrocarbon source. Initial sizes in Al_{7075} and Ag-C NP varied from 1-7 mm and 5-20 nm, respectively.

Composites were produced by mixing and milling Al_{7075} and Ag-C NP at different concentrations, from 0.5 to 1.5 wt. % of nanoparticles. An Al_{7075} without nanoparticles was produced as a reference material. For milling runs, a high energy shaker ball mill SPEX 8000M was used and the milling times were 5 and 10 h. The powders were consolidated by uniaxial load at 950 MPa for 2 minutes to generate a pellet of 4 cm in diameter, this pellet was sinterized first at 473 K for 2 h, then at 733 K for 4 h, then at 748 K for 18 h, and finally it was cooled down inside the furnace, all this into an argon atmosphere [9]. The sintered material was extruded at 773 K by the indirect extrusion where the pellet is forced to flow through a circular die; an extrusion relation of 25 was used under high pressure to produce bars of 10 mm in diameter and 500 mm of length.
The microstructural characterization was done by XRD in a Panalytical X′Pert PRO diffractometer, operated at 40 kV and 35 mA in the 2θ range of 20-100 degrees. Step and collection time were 0.05° and 5 s, respectively. High resolution transmission electron microscopy (HRTEM) studies were done in a JEOL 2200FS operated at 200 kV. The chemical analysis was carried out by characteristic X-ray energy dispersion spectroscopy (EDS) with a DX4 X-ray energy dispersive spectrometer coupled to the TEM. Mechanical properties were evaluated in a Shimadzu Universal Testing Machine with a load cell of 250 kN, the displacement rate of the crosshead was set at 0.066 mm/s, the samples were machined in accordance with the ASTM E8M standard.

Results

Microstructural characterization

Figures 1a and 1b show the structural evolution of the nanocomposites milled for 5 and 10 h, respectively. Diffraction peaks were broad and the spectra's full-width at half-maximum (FWHM) at higher concentrations of Ag-C NP were higher, denoting a crystal refinement due to the mechanical milling process [2, 3] and by the presence of silver nanoparticles [4]. Also, it can be appreciated smaller peaks of the $MgZn_2$ phase. Ag-C NP had an effect on the dissolution sequence of the $MgZn_2$ phase, because the characteristic reflections of this phase present a different intensity at different Ag-C NP concentrations [4].

Figures 2a and 2b show the structural evolution of the nanocomposites milled for 5 and 10 h, respectively, in the as-extruded condition. Diffraction peaks of the aluminum solid solution have a higher intensity compared with that of the figure 1, indicating a growth of the crystals during the sintered and hot extrusion processes. Beside this, it can be appreciated smaller peaks of different phases formed after the milling process: $MgZn_2$, Al_4C_3 and MgO. The $MgZn_2$ phase precipitates from the solid solution. It is supposed that the Al_4C_3 phase was formed of two different sources of carbon, the PCA and the Ag-C NP. Since the intensity of these peaks is higher with higher nanoparticles contents, these should be the principal source of carbon. The magnesium oxide is a very stable oxide in accordance with the Ellingham diagram [10], however, this phase is only present in the extrusion products. All these phases help in the strengthening of the nanocomposites.
Table I presents the crystallite size obtained by the Scherrer method in both, as-milled and as-extruded condition; milled products presents a random crystallite size values, extruded nanocomposites show a crystallite size reduction as a consequence of the increment of nanoparticles; which indicates that silver nanoparticles act like PCA, as it was mentioned before [4]. Comparing the both conditions, it is evident that the growth of crystallites is a consequence of sintering and hot extrusion processes.

Structural characterization

Figure 3 presents TEM images from the Al_{7075}–Ag-C NP nanocomposites. Figure 3a is a representative view from the nanocomposite, where a homogeneous distribution of silver nanoparticles into the matrix of aluminum is observed; these nanoparticles have a particle size between 5 and 80 nm. Figure 3b shows particles of different sizes into the matrix. A detailed analysis shows two different phases, dark particles corresponding to the Ag-C NP and grey particles which were identified as $MgZn_2$ phase. Figure 3c shows a close-up in an Ag-C NP, whose size is about 15 nm. It can be seen the spherical shape that silver nanoparticles conserve after the milling process.

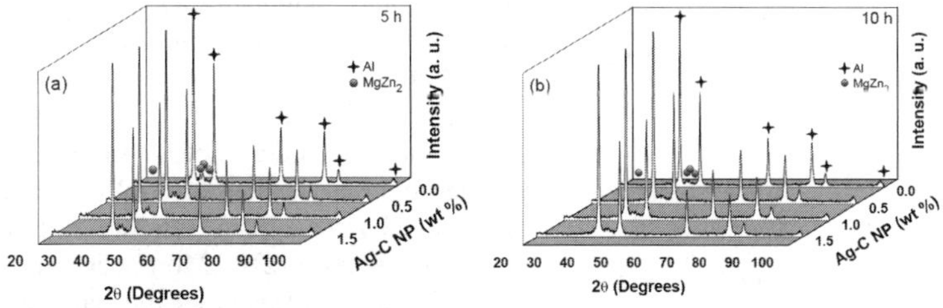

Figure 1. XRD patterns of the Al_{7075}–Ag-C NP nanocomposites processed by mechanical milling a) 5 h of milling, b) 10 h of milling.

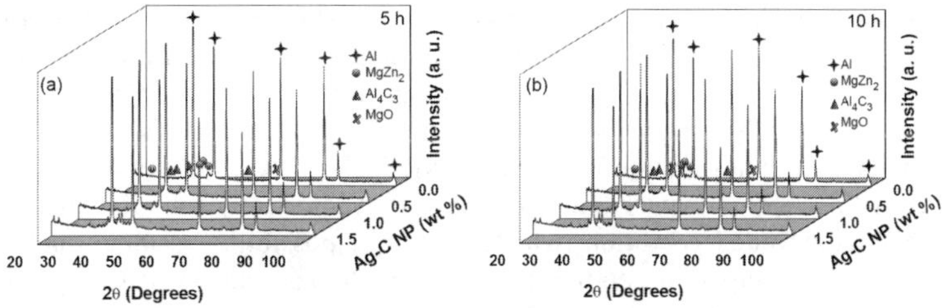

Figure 2. XRD patterns of the Al_{7075}–Ag-C NP nanocomposites processed by mechanical milling and hot extrusion. a) 5 h of milling, b) 10 h of milling.

Table I. Crystallite sizes as a function of Ag-C NP content, for the as-milled and as-extruded materials.

	Crystallite Size (Å)			
	As-milled		As-extruded	
Ag-C NP (wt. %)	**5 h**	**10 h**	**5 h**	**10 h**
0.0	533	565	889	919
0.5	628	476	886	914
1.0	544	449	820	819
1.5	565	408	803	793

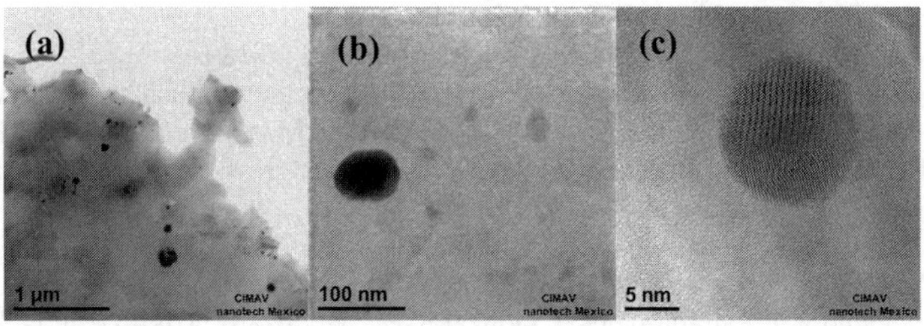

Figure 3. TEM micrographs of the Al_{7075}–Ag-C NP nanocomposite after 10 h of milling time and 1.5 wt. % silver nanoparticles, in the extruded condition.

Mechanical characterization

Figure 4 presents the yield strength (σ_y) and the ultimate tensile strength (UTS) as a function of the Ag-C NP content and the milling time (5 h and 10 h). It can be seen that the σ_y and UTS values for the Al_{7075} milled without nanoparticles are higher than those for the annealed pure Al (91.8 MPa)

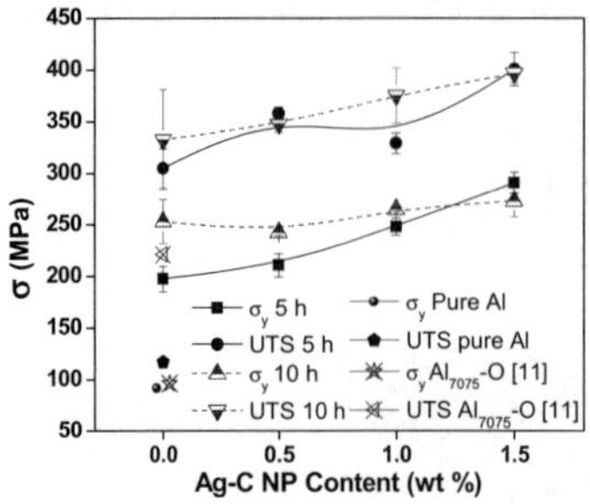

Figure 4. Yield strength σ_y and ultimate tensile strength (UTS) as a function of the milling time on the Al_{7075}–Ag-C NP nanocomposites, milled for 5 h and 10 h and hot extruded.

and Al_{7075} alloy (96.5 MPa) [11]. With additions of silver nanoparticles, these two properties are higher as the content of nanoparticles increases. σ_y and UTS values for 1.5 wt. % of Ag-C NP are ~ 275 MPa and ~ 400 MPa, respectively, as much for 5 as for 10 h of milling. The increment of the σ_y and UTS values is due to the mechanical deformation introduced in the powder by the mechanical milling, which produces materials with smaller crystallite size as the milling time increases [2]. Moreover the mechanical properties are increased as the nanoparticles content increases. This is due to two probable mechanisms: *i)* crystallite size reduction favored by the Ag nanoparticles, which act as a PCA [4], and *ii)* the inhibition of dislocation movements during the plastic deformation, caused by the second-phase particles that are homogenously dispersed into the matrix [8]. These particles are mainly Ag-C NP and those composed by MgO, Al_4C_3 and $MgZn_2$. By analyzing all figures, it is found that small additions of Ag-C NP (0.5-1.5 wt. %) could be used to produce nanocomposites with excellent mechanical properties. Additionally, all of these results are for composites in the as-extruded condition, which opens the possibility to obtain better mechanical properties after suitable tempering conditions (for example T6 temper). It is expected a synergic effect between Ag-C NP and the precipitated phase ($MgZn_2$) after aging treatments due to the dissolution effect that these particles present. However, to fully understand the contribution of Ag-C NP and precipitated phases to the strengthening in the Al-based nanocomposites, additional work is required.

Conclusions

Al based nanocomposites have been produced by dispersing Ag-C NP by mechanical milling and hot extrusion. The milling time and the Ag-C NP have an important effect on the particle and crystal size, as well as in the enhancement of mechanical properties. The excellent dispersion of the Ag-C NP into the Al_{7075} matrix, the precipitation of the $MgZn_2$ phase and the aluminum carbide formation, have also an important effect on mechanical properties. The best mechanical properties were found between 1.0 and 1.5 wt. % of silver nanoparticles.

References

[1] ASM Handbook, volume 21, Composites, 2001.
[2] C. Suryanarayana, *Progress in Materials Science 46* (2001) 1-184.
[3] L. Lu and M. O. Lai. *Mechanical Alloying*. (Kluwer Academic Publishers, Boston 1998).
[4] R. Flores-Campos, D. C. Mendoza-Ruiz, M. Martínez-Martínez, I. Estrada Guel, M. Miki-Yoshida, R. Martínez-Sánchez, XXIX International Congress on Metallurgy and Materials, Saltillo, Mexico (2007) 345-354.
[5] B. Prabhu, C. Suryanarayana, L. An, R. Vaidyanathan, Mater. Sci. Eng. A 425 (2006) 192-200.
[6] N. Zhao, P. Nash, X. Yang, J. Mater. Proc. Tech. 170 (2005) 586-592.
[7] M. I. Flores-Zamora, I. Estrada-Guel, J. González-Hernández, M. Miki-Yoshida, R. Martínez-Sánchez, J. Alloys Comp. 434-435 (2006) 518-521.
[8] G. E. Dieter. *Mechanical Metallurgy*. (Mc Graw Hill Book Company, Singapore 1988).
[9] T. Sheppard. *Extrusion of Aluminum Alloys*. (Kluwer Academic Publishers, Boston 1999).
[10] B. S. Mitchell, *An introduction to materials engineering and science: for chemical and materials engineers* (John Wiley & Sons Inc., New Jersey 2004).
[11] Information on http://www.matweb.com.

Materials Science Forum Vol. 644 (2010) pp 13-16

doi:10.4028/www.scientific.net/MSF.644.13

Synthesis of plasticizer-based ferrofluid and its use in the preparation of magnetic PVC nanocomposite

E. Servín-Hernández[a], O.S. Rodríguez-Fernández[b], L.A. García-Cerda[c]

Centro de Investigación en Química Aplicada (CIQA)
Boulevard Enrique Reyna Hermosillo No. 140, Saltillo, Coahuila, C. P. 25253, México

[a]elserher@hotmail.com, [b]oliverio@ciqa.mx, [c]lagarcia@ciqa.mx

Keywords: magnetite, ferrofluid, peptization method, magnetic nanocomposites, polymers.

Abstract: A magnetic polyvinylchloride (PVC) nanocomposite was prepared by static casting using a plasticizer-based ferrofluid. Two sets of nanocomposites were prepared: one under the influence of magnetic field and the other without magnetic field. The effects of ferrofluid content and the magnetic field on the magnetic and mechanical properties of the nanocomposite were studied in detail. Magnetite (Fe_3O_4) nanoparticles used for the ferrofluid preparation were prepared by the chemical co-precipitation method. Dioctyl phtalate (DOP)-based ferrofluid was prepared by the peptization technique. X-ray diffraction (XRD) was used to characterize the superparamagnetic behavior of the nanocomposite. A Universal tensometer was used to evaluate their mechanical properties. The results showed that the magnetization value of the nanocomposites increased as a function of ferrofluid concentration with all the samples showing superparamagnetic behavior. The mechanical studies showed that the tensile strength and elongation at break of the magnetic PVC nanocomposite were increased by the addition of ferrofluid and the applied magnetic field.

Introduction

Ferrofluids are magnetic liquids prepared as a stable colloidal solution consisting of monodomain magnetic nanoparticles dispersed in a non-magnetic carrier such as water, organic or inorganic liquid [1-4]. PVC is a useful polymer for industrial applications due to its good mechanical properties, good resistance to acidic and basic environments, fire retardancy and good processability at low cost [5]. There are different methods in the literature for the preparation of metallic-polymer composite. The simplest of all is by mixing monodomain magnetic particles with PVC [6]. Another method is by the synthesis in-situ of magnetic nanoparticles within a polymeric matrix [7].

Recently, our group reported the preparation of magnetic composite using a plasticizer-based ferrofluid [8]. In this work, our objective is to modify this procedure to prepare a DOP-based ferrofluid to be used for the preparation of magnetic PVC nanocomposite. A magnetic field during the preparation of the nanocomposite is used and its influence on the mechanical properties is investigated. The idea of using the magnetic field is to try to produce an alignment of both nanoparticles and polymer chains and to study the effect on mechanical properties.

Experimental Section:

Preparation of materials: Fe_3O_4 nanoparticles were prepared using a chemical co-precipitation method [6]. 24 g of $FeCl_3{\cdot}6H_2O$ and 12 g of $FeCl_2{\cdot}4H_2O$ ($FeCl_3$:$FeCl_2$, molar ration 2:1) were dissolved in 350 ml of deionized water under stirring at 70°C. 50 ml ammonium hydroxide (NH_4OH, 30% vol.) was then added quickly into the solution under vigorous stirring. A black Fe_3O_4 precipitate was obtained. A small quantity of this precipitate (for XRD characterization) was collected by magnetic decantation and washed five times with deionized water to remove any supernatant. The precipitate was finally dried at room temperature for its characterization. The rest of the precipitate in water was used for the ferrofluid preparation with DOP by the peptization technique [10] to consist of 15% w of Fe_3O_4 and 85% w of DOP. Dioctyl sodium sulfosuccinate (0.5 g) was added to the precipitate solution and stirred for 10 min. Then 50 ml of DOP was added and stirred for another 10 min. The obtained ferrofluid was collected by magnetic decantation. Any residual water was removed in a vacuum oven. Finally, 5 ml of polyethylene glycol octylphenyl ether was added under stirring.

Preparation of PVC nanocomposites: The magnetic PVC nanocomposites were formulated first by preparing a plastisol that consisted of 100 phr (parts per hundred) of PVC resin supplied by Primex with an average molecular weight of 54,000 g/mol, 140 phr of DOP (plasticizer) and 4 phr of Ca/Zn as the thermal stabilizer. To achieve a homogeneous mixture, all the components were stirred for 40 min. After that, the ferrofluid was added to the mixture with different concentrations of Fe_3O_4 (0.1, 0.5, 1, 2 and 5 phr) in it. The magnetic PVC nanocomposites were molded into 70x25x2 mm sheets by static casting of plastisol at 180°C for 3 min, followed by cooling to room temperature. Various sets of nanocomposites were prepared with and without magnetic field. Figure 1 shows the experimental setup used for the preparation of each nanocomposite under a longitudinal magnetic field (Neodymium permanent magnets, magnetic field of 11,700 to 12,100 Gauss).

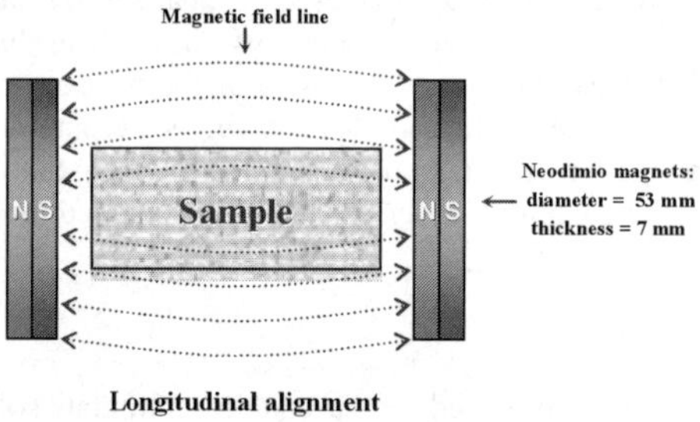

Fig. 1. Experimental setup for the magnetic applied field.

Characterization: XRD pattern was obtained with a Siemens diffractometer (D-5000) at room temperature using CuK_α (25 mA, 35 kV). The hysteresis loop was recorded in a SQUID magnetometer (PPMS Quantum Design model 6000) at room temperature, under a maximum applied field of 20 kOe. Tensile tests were performed on a Universal machine CEF-80 tensometer according to ASTM-D638 at 21 °C and 50 % of relativity humidity. Crosshead speed was set up at 50 mm/min. The samples for the test were obtained in the direction of magnetic field. Each data point was the average of five to seven samples.

Results and Discussion

XRD pattern for the magnetite obtained by the co-precipitation method is shown in Figure 2a. The main reflections from planes (220), (311), (400), (511) and (440), characterizing spinel ferrites are strongly observed. The reflections from the other planes (111), (422) and (533) can be observed but with weak intensity. The analysis of XRD pattern indicates that the Fe_3O_4 sample has a spinel cubic structure with a single phase [11]. The magnetite nanoparticles used in this work have an average size of ~ 12 nm, calculated from the Scherrer´s formula [12].

Figure 2b shows the hysteresis loop of Fe_3O_4. The magnetization (M) versus applied field (H) plot for the magnetite shows that the saturation magnetization (M_S), coercivity (H_C) and remanent magnetization (M_R) are 74.2 emu/g, 65 Oe and 4.7 emu/g, respectively, indicating that the nanoparticle has a superparamagnetic behavior.

The hysteresis loop for the DOP-based ferrofluid is shown in Figure 3a in which a superparamagnetic behavior is also observed and the values for M_S, H_C and M_R are 10.4 emu/g, 10 Oe and 0.13 emu/g, respectively. Figure 3b shows the hysteresis loops for PVC and the magnetic PVC nanocomposite with 0.5 phr Fe_3O_4 in the ferrofluid. PVC has a diamagnetic behavior while the magnetic nanocomposite shows a superparamagnetic behavior with a magnetization value of 0.126 emu/g.

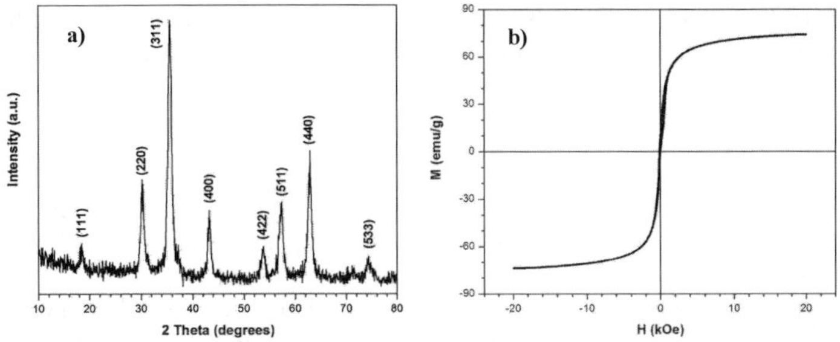

Fig. 2. X-ray diffraction pattern (a) and hysteresis loop (b) for the magnetite.

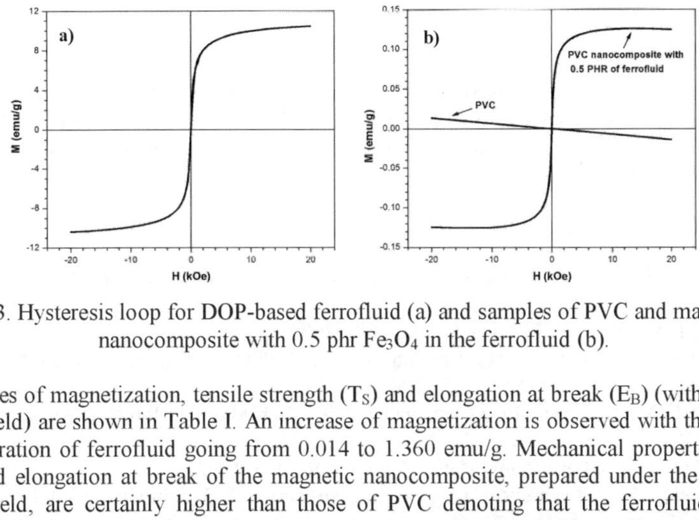

Fig. 3. Hysteresis loop for DOP-based ferrofluid (a) and samples of PVC and magnetic nanocomposite with 0.5 phr Fe_3O_4 in the ferrofluid (b).

The values of magnetization, tensile strength (T_S) and elongation at break (E_B) (with and without magnetic field) are shown in Table I. An increase of magnetization is observed with the increase in the concentration of ferrofluid going from 0.014 to 1.360 emu/g. Mechanical properties of tensile strength and elongation at break of the magnetic nanocomposite, prepared under the influence of magnetic field, are certainly higher than those of PVC denoting that the ferrofluid acts as an effective reinforcing agent in improving these properties as also reported in similar works [13, 14].

		With field		Without field	
Nanocomposite	**M_S (emu/g)**	**T_S (MPa)**	**E_B (%)**	**T_S (MPa)**	**E_B (%)**
PVC	Diamagnetic	3265	347	3265	347
PVC + 0.1 phr of Fe_3O_4	0.014	4125	453	1490	116
PVC + 0.5 phr of Fe_3O_4	0.126	4158	415	3356	349
PVC + 1 phr of Fe_3O_4	0.255	4291	403	3889	401
PVC + 2 phr of Fe_3O_4	0.573	4182	383	1932	159
PVC + 5 phr of Fe_3O_4	1.360	3980	365	1600	105

Table I. The magnetic and mechanical properties for PVC and the magnetic nanocomposite with and without magnetic field.

Very interestingly, we can see that mechanical properties of the nanocomposites prepared with and without magnetic field are inconsistent. The samples prepared with 1 phr of Fe_3O_4 show the best T_S and E_B, while further addition of more Fe_3O_4 would cause a reduction in both properties. Our results suggest the maximum addition of Fe_3O_4 should not be more than 1 phr. At this amount, T_S and E_B are increased significantly by 31 % and 16%, respectively for the nanocomposites treated with magnetic field. This enhanced behavior can be interpreted in terms that magnetic field helps to align the magnetic nanoparticles in the nanocomposite to obtain an anisotropic material as has been reported in similar work [15]. The unfavorable effect of using more than 1 phr could be attributed to agglomerates formed as a result of the saturation of Fe_3O_4 in the nanocomposites.

Summary

DOP-based ferrofluid was successfully prepared by the peptization method in our work. This ferrofluid was also successfully used to prepare a magnetic PVC nanocomposite, with and without magnetic field. The nanocomposite with magnetic field showed a superparamagnetic behavior with magnetization values in the range of 0.014 to 1.360 emu/g. The mechanical properties of the nanocomposite with magnetic field were improved also. Our procedure can be used to further obtain other ferrofluid-nanoparticles systems for preparing magnetic nanocomposites with different polymers.

Acknowledgment

This work was supported by CONACyT-México under grants 46302 and 46044. E. Servín thanks CONACyT for the scholarship No. 173273. The authors would like to thank Hilda Lucia Prado López, Gilberto Hurtado López and Sion S. Ng for their help in running tensile test, VSM measurements and commentaries to the writing.

References

[1] B. Berkovski, V. Bashtovoy, Magnetic Fluids and Applications Handbook, UNESCO Series of Learning Materials, New York, NY, 1996.
[2] A.Y. Zubarev, L.Y. Iskakova, Physica A, Vol. 367, (2006), p 55.
[3] C. Scherer and A.M. Figueiredo Neto, Brazilian Journal of Physics, Vol. 35, 3A, (2005), p. 718.
[4] S. Odenbach, Colloids Surf., A, Vol. 217, (2003), p. 171.
[5] L.I. Nass and C.A. Heiberger, Encyclopedia of PVC, 2nd edition, Marcel Dekker, Vol. 1-4, (1998).
[6] L.L. Beecroft and C.K. Ober, Chem. Mater., Vol. 9, (1997), p. 1302.
[7] L.A. García-Cerda, M.U. Escareño-Castro, M. Salazar-Zertuche, J. Non-Cryst. Solids, Vol. 353 (2007), p. 808.
[8] I. G. Yáñez-Flores, R. Betancourt-Galindo, J. A. Matutes Aquino, O. S. Rodríguez-Fernández, J. Non-Cryst. Solids, Vol. 353, (2007), p. 799.
[9] R. Betancourt-Galindo, R. Saldívar-Guerrero, O. S. Rodríguez-Fernández, L. A. García-Cerda, J. A. Matutes Aquino, J. Alloy. Comp., Vol. 369, (2004), p. 87.
[10] S. E. Khalafalla and G. W. Reimers, IEEE Transactions on Magnetics, Vol. 16, (1980), p. 178.
[11] Joint Committee on Powder Diffraction Standard (JCPDS), Card No. 19-0629.
[12] B. D. Cullity and S. R. Stock: Elements of X-ray Diffraction, 2nd edition, Ed. Addison-Wesley, (1978), USA.
[13] L. Zhang, X. Chen, C. Li, J. Mater. Sci., Vol. 40, (2005), p. 2097.
[14] J. Alam, U. Riaz. S. Ahmad, J. Magn. Magn. Mater., Vol. 314, (2007), p. 93.
[15] D. Shi, P. He, J. Lian, X. Chaut, S. L. Budko, E. Beaugnon, L. M. Wang, R. C. Ewing, R. Tournier, J. Appl. Phys., Vol. 97, (2005), p. 06312.

Materials Science Forum Vol. 644 (2010) pp 17-20
© (2010) Trans Tech Publications, Switzerland
doi:10.4028/www.scientific.net/MSF.644.17

EFFECT OF TYPE AND CONCENTRATION OF IONOMER COMPATIBILIZER ON THE HDPE/ IONOMER/ CLAY NANOCOMPOSITES MORPHOLOGY

S. López García[a], S. Sánchez-Valdés[b] and L.F. Ramos-deValle[c]
Centro de Investigación en Química Aplicada
P.O. Box 379, Saltillo, Coahuila, CP-25253, México
[a]lopezgs@posgrado.ciqa.mx, [b]saul@ciqa.mx, [c]devalle@ciqa.mx

Keywords: ionomer compatibilizer, morphology, high density polyethylene nanocomposites.

ABSTRACT: A study of the effect of an ionomer compatibilizer (surlyn® 9520 and 9721, both with zinc as the neutralizing cation) on the morphology and properties of a high density polyethylene (HDPE) - montmorillonite clay nanocomposite was carried out. The nanoclay used was cloisite 20A®. Polyethylene /Ionomer /Clay nanocomposites were prepared via melt mixing in a twin screw extruder. The nanoclay dispersion and exfoliation were examined through X-Ray Difraction (XRD) and Scanning Electron Microscopy (SEM). TGA was carried out to determine the effect of nanoclay on the thermal stability of the HDPE nanocomposites. Results showed that both ionomers impart a marked compatibility between the polymer and the nanoclay, promoting the exfoliation of the nanoclay within the HDPE matrix. Nonetheless, ionomer 9520 (with the higher degree of neutralization) at 10 and 12 wt% content produced completely exfoliated morphologies, whereas, the ionomer 9721 produced a lesser degree of exfoliation with few tactoids.

Introduction

Polymeric nanocomposites represent a new class of materials which arise from the dispersion of nanometer size particles in a polymer matrix. The properties of these nanocomposites depend largely on the polymer - nanoparticles interaction and on the dispersion of these nanoparticles. When the nanoparticles are clays of the montmorillonite type, those properties depend also on the degree of exfoliation[a] of the clay particles. With respect to the mentioned interactions, polarity plays a very important role. Strong interactions are promoted when both, polymer and nanoparticles, are either hydrophilic or hydrophobic [1,2]. In the case of polyethylene (hydrophobic non-polar) - montmorillonite clay (hydrophilic polar) nanocomposites, the polymer and/or the clay should have to be modified in order to promote those interactions. In addition, a compatibilizer can also be used to further promote the interactions in order to obtain an exfoliated polyethylene-montmorillonite nanocomposite [3].

[a] *Exfoliation is the process by which the larger clay particles called tactoids, of 1-10 μm in size, comprised by hundreds of platelets or laminates stacked together by strong attraction forces, become delaminated or exfoliated into hundreds of clay nanoparticles of 10-100 nm.*

Experimental Section

High density polyethylene PADMEX-65050 from Pemex, México, with an MFI of 5.0 g/10 min was used. Two different Surlyn ionomers (9520 and 9721) from DuPont, USA, with an MFI of 1.1 and 1.0 g/10 min, respectively, both with zinc as the neutralizing cation, and 9520 with a higher degree of neutralization, were used. The montmorillonite clay used was cloisite 20A from Southern Clay Products, USA (*http://www.scprod.com*), modified with 95 meq of a quaternary ammonium salt /100 g of clay.

Nanocomposite Preparation

Polyethylene /Ionomer /Clay nanocomposites were prepared via melt mixing in a Werner & Pfleiderer ZSK30 twin screw extruder, using a relatively high shear stress screw configuration, at 20 rpm. The temperature profile was from 180 °C at the feed zone to 200 °C at the die zone. Three different ionomer concentrations used were, 8, 10 and 12 wt%, whereas, the clay concentration was kept constant at 4 wt%.

Evaluation of Nanocomposites

X-ray diffraction of clay and nanocomposites, in order to evaluate the exfoliation of the montmorillonite clay was performed in a Siemens D5000 (25 mA, 35 kV) using CuKα X-ray radiation, at 0.6°/min, from 1-10°. Samples for X-ray analysis were obtained from compression molded plaques. The thermal properties were investigated using a 951 Thermogravimetric analyzer from Dupont Instruments. The samples were examined at a heating rate of 10°C/min under a nitrogen flow, from 25 to 650 °C. To study the distribution and dispersion of the nanoclay and to support the XRD analysis on exfoliation, Field Emission Scanning Electron Microscopy (FESEM) was carried out on ultrathin sections approximately 70-100 nm in thickness, cut from the compression molded plaques, with a diamond knife using a Leica microtome. The FESEM observations were performed with a JSM-7401F electron microscope.

Results and Discussion

Figure 1 shows the XRD patterns of the cloisite 20A clay and the HDPE/ Clay nanocomposites with and without ionomer, at 8, 10 and 12 wt% ionomer content. It is observed that the basic reflection peak of the cloisite 20A clay appears at 3.7 ° which, according to Bragg's law, would correspond to an intergallery spacing of 23.9 Å (*i.e. the spacing between the stacked laminates that form the tactoid*). The nanocomposite without ionomer shows exactly the same reflection peak and thus, the same inter-gallery spacing as the sole cloisite 20A.

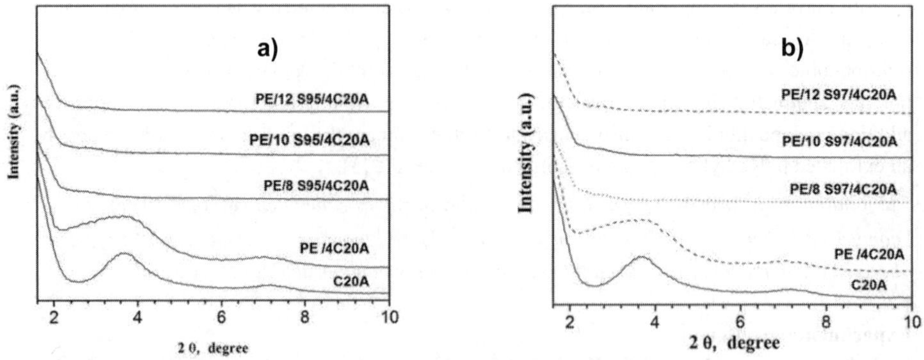

Figure 1. XRD difractograms of the sole cloisite 20A clay, the HDPE/ Clay nanocomposite without ionomer, and the HDPE/ Ionomer/ Clay different nanocomposites; 1-a, with ionomer 9520 and 1-b, with 9721. a) b)

On the contrary, all HDPE/ Ionomer/ Clay nanocomposites show no reflection peak, which according to Vaia et al [4] would suggest a high level of exfoliation of the clay platelets. Those authors propose that the increase in the intergallery spacing is associated with the intercalation of the polymer molecules within the clay galleries, resulting in a new diffraction pattern that would correspond to the expanded inter-gallery spacing in the nanocomposite. Thus, the absence of any XRD peak suggests a considerable separation between the clay platelets, which would imply a more extensive interaction between the polymer and the surface of those clay platelets.

Table 1 presents the temperature at which pure HDPE and the nanocomposites reach 50 wt% decomposition. First, it is observed that the nanocomposites show a higher decomposition temperature that pure HDPE; and second, the nanocomposites with ionomer 9520, which has the higher degree of neutralization, show a significantly higher decomposition temperature than those with 9721. This increase in the decomposition temperature can be related with the presence of clay, but the difference between the two series of nanocomposites can be related with a slightly better intercalation-exfoliation of the clay platelets in the nanocomposites using the ionomer with the greater level of neutralization, that is, I-9520.

		Ionomer 9520	Ionomer 9721
Pure HDPE	-----	482 °C	482 °C
HDPE/ Clay/ Ionomer samples, all with **4 wt%** clay	**8** wt% Ionomer	485	483
	10 wt% Ionomer	488	485
	12 wt% Ionomer	497	486

Table 1. Temperatures at which there occurs 50 wt% decomposition of HDPE and the Nanocomposites, from TGA.

Figure 2 illustrates three different types of morphologies presented by the prepared nanocomposites. Figure 2a corresponds to the HDPE/ **4Clay** nanocomposite, without ionomer, where a very poor dispersion can be appreciated, which is assumed to be due to the incompatibility between the polymer and the clay. Figures 2b and 2c correspond to the HDPE/ Ionomer-9520/ **4Clay** nanocomposites with 8 and 10 wt% ionomer content, respectively. A much better dispersion can be observed, due to the presence of I-9520. Also, comparing Figures 2b and 2c, it appears that the level of exfoliation increases as the content of ionomer I-9520 increases from 8 to 10 wt%, showing the effect of the ionomeric compatibilizer on the nanocomposite morphology. Figure 2d, on the other hand, corresponds to the HDPE/ Ionomer 9721/ **4Clay** nanocomposite, where there appear tactoids in slightly exfoliated zones. This emphasizes the difference between the two different ionomers, with the 9520 (the one with the higher degree of neutralization) producing the greater compatibility, and thus the higher lever of exfoliation in the HDPE/ clay nanocomposites.

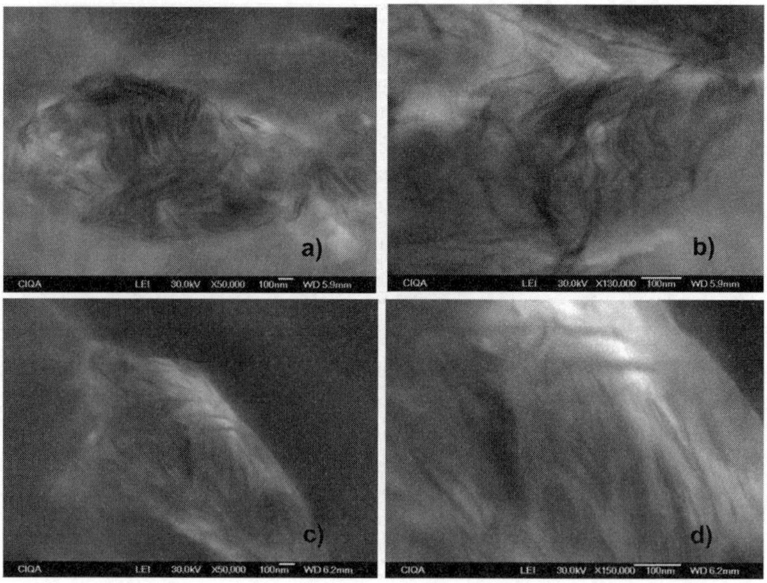

Figure2. SEM micrographs in STEM mode of: a) HDPE/ Clay; b) HDPE/ 8%-I9520/ Clay; c) HDPE/ 10%-I9520/ Clay; and d) HDPE/ 10%-I9721/ Clay.

Conclusions

HDPE/ Ionomer/ Clay nanocomposites were prepared via melt mixing. The use of ionomer greatly improved the compatibility between HDPE and montmorillonite clay and greatly favored the exfoliation of the clay. XRD showed insufficient to establish the exfoliation of the clay. However, XRD in combination with electron microscopy, showed to be much more accurate in determining this exfoliation. With respect to the effect of the ionomer type, it appears that the higher the level of neutralization, the greater the compatibilizing effect between HDPE and clay, and thus the greater the effect towards favoring the exfoliation of the clay. The decomposition temperature of HDPE increased with the addition of clay. However, this increase was more pronounced when using the ionomer with the higher neutralization level, that is, I-9520; which could be attributed to the higher level of exfoliation in these nanocomposites.

Acknowledgments

The authors thank CONACYT for its financial support to carry out this study, through project 84424. One of the authors, S.L.G., thanks CONACYT for a scholarship to carry out her PhD studies. The authors also thank M. Sanchez-Adame, M. C. Gonzalez-Cantu, T. Rodriguez-Hernandez, J. Rodriguez-Velazquez, M. G. Mendez-Padilla, E. Hurtado-Suarez, P. Siller-Flores and Daniel A. Alvarado for their technical and information support.

References

[1] M. Alexandre and P. Dubois; Materials Science and Engineering, (2000), 28, 1.
[2] R. S. Sinha and M. Okamoto; Progress in Polymer Science, (2003), 28, 1539.
[3] E. Manias, A. Touny, L. Wu, B. Lu, K. Strawhecker, J. Guilman, T. Chung; Polym. Mater. Sci. Eng. (2000), 82, 282.
[4] R. A. Vaia and E. P. Giannelis; Macromolecules, (1997), 30, 7990.

Materials Science Forum Vol. 644 (2010) pp 21-24
doi:10.4028/www.scientific.net/MSF.644.21

Elemental analysis of a heterogeneous polymeric system by EDS: Detection of the compatibilizer agent containing Si atoms and silver nano-particles (AgNP´s) in High Impact Polystyrene

F. Soriano[1,a], C. de Anda[1,b], G. Morales[1,c], A. Ponce[1,d]

[1]Research Center of Applied Chemistry, Saltillo, Mexico 25253

[a]fsoriano@ciqa.mx, [b]candavi@posgrado.ciqa.mx, [c]gmorales@ciqa.mx, [d]aponce@ciqa.mx

Keywords: EDS analysis, high impact polystyrene, heterogeneous polymeric system

Abstract. High impact polystyrene (HIPS) is considered a heterogeneous polymeric system and is constituted by a disperse phase of rubber particles within a continuous phase formed by a polystyrene (PS) matrix. During the synthesis of HIPS, PS is formed whereas some PS chains become chemically bonded to polybutadiene (PB) chains, resulting in a PB-*g*-PS copolymer, which decreases the interfacial tension between both incompatible phases, acting as a compatibilizer. In this study, 2 HIPS systems were evaluated by energy dispersive spectroscopy (EDS). In the first case, a HIPS composite was analyzed by EDS to locate the graft copolymer species containing Si atoms, which were previously synthesized by anionic polymerization using chloro-silane chemistry. The graft copolymer was incorporated to the HIPS synthesis from the beginning of the reaction in order to provide an improvement on the rubber phase stability. The graft copolymer was located at the interphase between the PS occlusions and the PB subdomains of the rubber particles. In the second case, HIPS was synthesized incorporating silver nanoparticles (AgNP´s) during the polymerization reaction, where the system can exhibit different types of morphology of the elastomeric phase (micelles, lamellas and core-shell). These structures overlapped the AgNP´s in the transmission electron microscopy (TEM) images, thus resulting this technique to be inadequate to establish the location of the AgNP´s. In this context EDS was used to evaluate the location and distribution of the AgNP´s by means of the elemental composition analysis. AgNP´s were preferentially detected in the PS phase as bigger clusters.

Introduction

High impact polystyrene (HIPS) combines the rigidity of the polystyrene with the toughness of the polybutadiene, resulting in a material with a broad range of applications. The improvements in the mechanical properties of the HIPS, especially in impact strength, are attributed to the presence of rubber particles dispersed within a PS matrix. Different types of morphologies can be obtained by modifying the interfacial parameters between the disperse rubber phase and the continuous PS phase [1].

The graft copolymer (PB-*g*-PS) formed in situ during the synthesis of HIPS has an essential role on the development of the rubber particles. It has been reported that depending on the molecular characteristics of the graft copolymer (number of PS graft per PB chain and molecular weight of the PS graft), it can be located 1) in the interface between the rubber particles and the PS matrix or 2) it can be stabilizing the PS occlusion inside the rubber phase [2].

In the first case, this study involves the synthesis of HIPS using a model graft copolymer containing Si atoms, which were previously synthesized by anionic polymerization using chloro-silane chemistry. The second case, involves the synthesis of HIPS using a styrene-butadiene copolymer (SB) with a PB/PS = 70/30 composition where silver nano particles (AgNP´s) were incorporated into the reaction. Both materials were characterized by transmission electron microscopy (TEM) and energy dispersive spectroscopy (EDS) in order to provide a better understanding of the role of the graft copolymer on the formation of the rubber phase (first case), and to elucidate the distribution and location of AgNP´s in each phase (second case).

Experimental part

Materials. Styrene monomer (Poliformas Plásticas, México), PB rubber (M_n = 100,000 g/mol, Total Petrochemicals) and SB copolymer with a PB/PS = 70/30 wt-% composition, M_n = 267,000 g/mol, Dynasol Elastómeros S.A de C.V.) were used as received. AIBN (Sigma-Aldrich) was crystallized from ethanol. Benzoyl peroxide (BPO) (Sigma-Aldrich) was crystallized from methanol. The model graft copolymer with 1 PS graft ($M_{n\ PB\ backbone}$ = 72,000 g/mol and $M_{n\ PS\ graft}$ = 115,000 g/mol) was synthesized by high vacuum anionic polymerization using chloro-silane chemistry as described elsewhere [3]. Styrene/AgNP´s dispersion was supplied by Servicios Industriales Peñoles S. A. de C.V. The AgNP´s present a bimodal size distribution with average particle diameter (D_p) of 4 and 28 nm.

Instrumentation. TEM analysis was performed in a FEI-Titan 80-300 kV field emission gun microscope operated at 300 kV. For the chemical quantitative analysis, EDS has been used in the Z-contrast imaging supporting feature with the high angle annular dark field (HAADF) detector in STEM operation mode. For the TEM-EDS analysis, the specimens have been prepared using ultramicrotomy method for cutting electron-transparency slices.

HIPS synthesis. In the first case, HIPS was synthesized using reaction vessels of 15 mL capacity at 70 ºC and 250 rpm. The synthesis was carried out by bulk radical polymerization using 6% w/w of total rubber, where 7% of the total rubber corresponds to the PB from the model graft copolymer. AIBN (0.1 % w/w) was used as the radical initiator, since the grafting reaction is minimized in comparison with other radical initiators [4]. In the second case, the synthesis of HIPS was carried out using a gallon capacity stainless steel reactor with an anchor-turbine stirrer: i) the dissolution stage was carried at room temperature and the stirring rate was fixed at 40 rpm during 12 h. All reactants were added (see Table 1) except the BPO, ii) BPO was added and the stirred bulk pre-polymerization stage took place at 40 rpm and 90°C until monomer conversion (X) close to 0.35-0.40 (t = 225 min), several samples was extracted during polymerization at different predetermined periods of time and were monomer extracted.

Table 1. Recipe used for HIPS synthesis (Case 2).

Reactant	SB	Styrene	AgNP´s	BPO
Wt-%	8	91.8	0.1	0.1

Results

The HIPS obtained following the experimental conditions previously described (case 1) presented a salami type morphology (Fig. 1-A), with an average particle diameter (D_p) of 1.54 μm. Energy dispersive x-ray (EDS) analysis of the resulting sample provided further information on the localization of the graft copolymer due to the detection of the Si atom. In Fig. 1-B, it is shown the EDS analysis obtained for different locations within the HIPS morphology.

From Fig. 1-B1 and 1-B2, it can be observed that the graft copolymer is located at the interface of the small occlusions. The bigger occlusions of the rubber particles are formed by the coalescence of smaller ones, since they are not stabilized by the graft copolymer. Therefore, the small occlusions where the graft copolymer is present do not coalesce due to the integrity given by a reduced interfacial tension given by the graft copolymer.

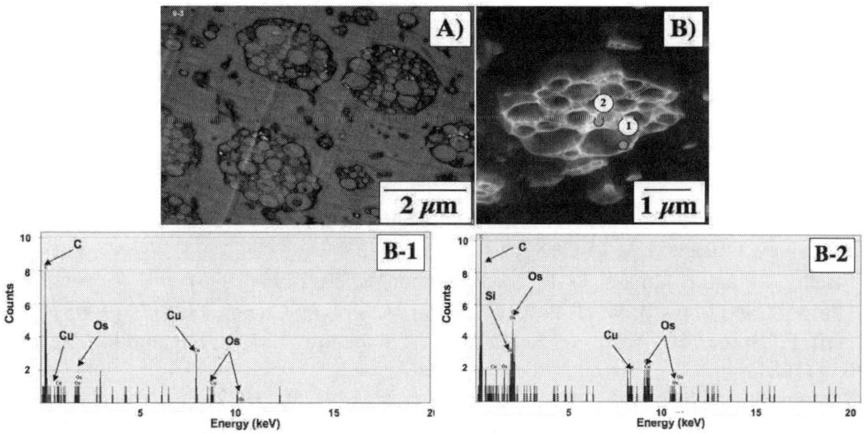

Fig. 1. A) TEM micrograph of HIPS and B) EDS analysis of HIPS (Case 1)

In Fig. 2 TEM micrograph and EDS analysis corresponding to HIPS samples (period of co-continuity) are shown (Case 2). Before phase inversion, a lamellar morphology can be observed (white area in the micrograph), which is characteristic of the rubber used [5, 6]. Also, agglomerated AgNP´s (primary D_p= 4 and 28 nm) can be observed indicating a poor dispersion probably as a consequence of the surfactant removal under the reaction conditions previously described, which was evidenced in the elastomeric phase as N (a chemical component of the surfactant) was detected by EDS analysis (Fig. 2-a3).

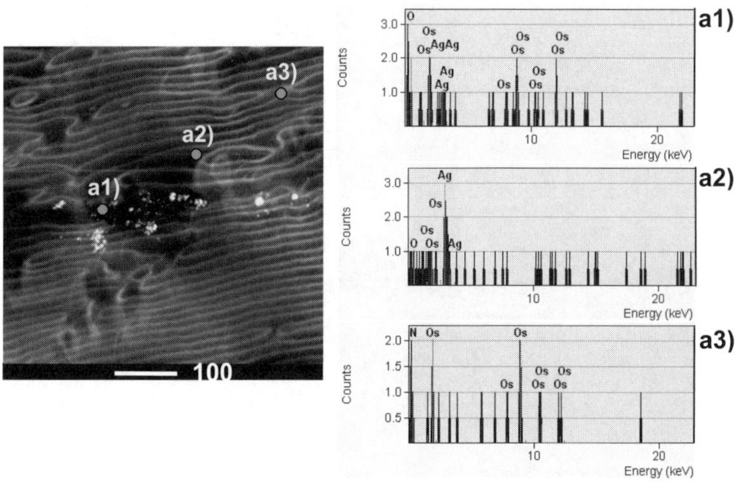

Fig. 2. TEM micrograph and EDS analysis, [AgNP´s] = 0.1 wt-%

With respect to the location of AgNP´s, TEM micrograph and the EDS spectra (Fig. 2- a2) show that AgNP´s was located preferentially in the PS phase (dark area in the TEM micrograph). Moreover, pure Ag as well as oxidized silver nano-particles were detected by EDS (Fig. 2-a1).

Conclusions

The model graft copolymer with 1 PS graft could be located at the interface of the small occlusion in the salami morphology resulting in a decrease of the interfacial tension, thus preventing the coalescence to form bigger occlusions. With respect to the AgNP´s, they were detected in the PS phase as bigger clusters as a consequence of the surfactant lost.

References

[1] Echte, A., Rubber-Toughened Plastics (C. Keith Riew Ed.), Proc. 194th National Meeting of ACS, New Orleans, LA, 15 (1987)
[2] Fischer, M. and Hellmann, G. P., Macromolecules, 29, 2498 (1996)
[3] Uhrig, D. and Mays, J. W., J. Polym. Sci. Part A: Polym. Chem., 43, 6179 (2005)
[4] Haff, F., Breuer, H., Achte, A., Schmitt, B. J., Stabenow, J., J. of. Sci. and Ind. Research, 40, 659 (1981)
[5] Sardelis H., Michels H. J., Allen G., Polymer, 28, 244-250 (1987)
[6] Díaz de León R., Morales G., Polymer Engineering and Science, 47 (11), 1827-1838 (2007)

Materials Science Forum Vol. 644 (2010) pp 25-28
© (2010) Trans Tech Publications, Switzerland
doi:10.4028/www.scientific.net/MSF.644.25

Dielectric properties of PMMA-SiO_2 hybrid films

M.D. Morales-Acosta[1,a], M.A. Quevedo-Lopez[2], H. N. Alshareef[3], B. Gnade[2] and R. Ramirez-Bon[1,b]

[1]CINVESTAV, Libram. Norponiente 2000 Fracc. Real de Juriquilla, Querétaro 76230, Qro., México

[2]Material Science and Engineering, The University of Texas at Dallas, Richardson 75083, TX, USA

[3]King Abdullah University of Science and Technology, Jeddah, Saudi Arabia

[a] mdmorales@qro.cinvestav.mx, [b] rrbon@qro.cinvestav.mx

Keywords: Hybrid PMMA-SiO_2; organic-inorganic materials; hybrid gate dielectric; flexible electronics; capacitor

Abstract. Organic-inorganic hybrid films were synthesized by a modified sol-gel process. PMMA-SiO_2 films were prepared using methylmethacrylate (MMA), tetraethil-orthosilicate (TEOS) as silicon dioxide source, and 3-trimetoxi-silil-propil-methacrylate (TMSPM) as coupling agent. FTIR measurements were performed on the hybrid films to confirm the presence of PMMA-SiO_2 bonding. In addition, metal-insulator-metal (MIM) devices were fabricated to study the dielectric constant of the films as function of frequency (1 KHz to 1 MHz). Electrical results show a weak trend of the dielectric constant of the hybrid films with MMA molar ratio. More importantly, the PMMA-SiO_2 hybrid films showed a higher dielectric constant than SiO_2 and PMMA layers, which is likely due to the presence of additional C-O-C bond.

Introduction

The combination of organic and inorganic components at the molecular scale in advanced hybrid materials provides new alternatives to achieve novel materials with specific properties [1-3]. The sol-gel method is the most convenient and suitable method to get the coupling between organic and inorganic phases at the interface level. Some of the advantages of this method include low temperature processing as well as high control in the final organic to inorganic phase composition. These advantages are of special interest in the fabrication of flexible electronic devices where the processing temperature is limited by the low processing temperature tolerated by the substrate. SiO_2-PMMA is one of the most intensively studied hybrid systems in the recent years. Highly transparent PMMA-SiO_2 films with different organic to inorganic molar ratios have been obtained by the sol-gel process [4-5]. The properties of the hybrid PMMA-SiO_2 films depend on their composition. Therefore some properties such as hardness, refraction index, etc. can be tuned by choosing the appropriated initial compositions of the precursors. In this paper, we report the dielectric properties of SiO_2-PMMA hybrid films. The hybrid films were synthesized varying the MMA molar ratio in the sol-gel precursor solution. In particular, we report the effect of the MMA molar ratio composition on the dielectric properties of the hybrid films.

Experimental details

PMMA-SiO_2 hybrid films. The hybrid PMMA-SiO_2 coatings were prepared by a modified sol-gel process using methylmethacrylate (MMA), tetraethil-orthosilicate (TEOS) as silicon dioxide source, and 3-trimetoxi-silil-propil-methacrylate (TMSPM) as coupling agent [1]. The SiO_2-PMMA films were prepared from solutions using 1.0:1.0 molar ratios for TEOS: TMPSM and different MMA molar ratios (0.25, 1.0 and 2.0). The starting solution was spin coated onto cleaned substrates and cured in a conventional oven at 90°C for 6 hrs.

Fabrication of the MIM capacitors. Metal-insulator-metal (MIM) devices were fabricated on p-type, (100)-oriented silicon wafers. The silicon substrates were chemically cleaned by standard RCA process and subsequent etching in diluted HF solution to remove the native SiO_2 layer on Si. In sequence, 10 nm of chromium and 100 nm of gold were thermally evaporated. After that, PMMA-SiO_2 films were deposited by spin coating with the conditions previously described.

Finally, the metallic gold contacts for the capacitor were also thermally evaporated through a shadow mask. The shadow mask consists of several areas (50, 100, 200, 300 and 500 μm diameters).

Characterization. Fourier Transformed Infrared (FTIR) spectroscopy absorption measurements were carried out using a Nicolet 6700 spectrometer in Grazing Angle Attenuated Total Reflectance (GA-ATR) mode with a resolution of 4 cm^{-1}. A background spectrum was subtracted from the measurements and was taken using a reference sample consisting of p-type (100) Si wafer cleaned and with 100 nm thickness of thermally evaporated gold layer. FTIR was carried out in the spectral region between 4000 to 650 cm^{-1}. The determination of the peaks was done by comparison with reference spectra of known materials.

SEM cross-section analysis of the PMMA-SiO_2 films was performed in a Zeiss Supra-40 SEM equipment operated at 10 kV, to evaluate the dielectric thickness of each sample. Before measurement, the samples were coated with gold.

C-V measurements were carried out in an Agilent 4284A parameter analyzer covering a range from 1 to 1,000 KHz. The dielectric constant (k_i) value was calculated from the measured capacitance (C) using the equation:

$$C = k_i \varepsilon_0 \frac{A}{\chi_o} \quad (1)$$

Where ε_0 is the permittivity of free spaces, χ_o is the dielectric thickness and A is the capacitor area.

Results and discussion

Fig. 1 (a) shows the normalized FTIR spectra for the PMMA-SiO_2 hybrid films with a TEOS:TMSPM ratio of 1:0.25 and MMA molar ratios of 0.25, 1.0 and 2.0. The band assignments for PMMA-SiO_2 have already been reported [5]. The absorption band centered at 940 cm^{-1} is attributed to the Si-O-Si stretching vibration. The wide band centered about 1100 cm^{-1} is the result of an overlap band effect, it corresponds to Si-O (1060 cm^{-1}), Si-O-C (1130 cm^{-1}) and C-O-C (1200 cm^{-1}) bonds. The presence of the characteristics PMMA and SiO_2 bands shows a successful synthesis of the hybrid PMMA-SiO_2 system. There is no clear impact in the resulting PMMA-SiO_2 bonding characteristics as result of the different MMA ratios. For comparison, we show in Fig. 1(b) the impact of TMSPM concentration in the bonding characteristics of the resulting PMMA-SiO_2 hybrid; decreasing the TMSPM molar ratio results in a clear change in the broad band centered at about 1100 cm^{-1}. As mentioned before, this band is composed of Si-O, Si-O-C and C-O-C.

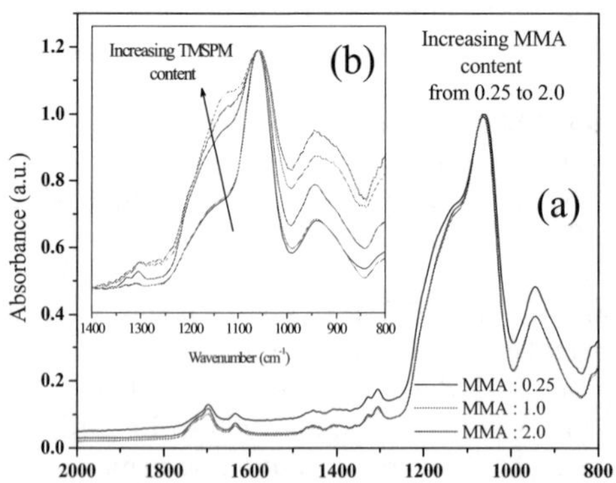

Fig. 1. (a) FTIR spectra of the PMMA-SiO_2 hybrid layer using different MMA ratios and a constant TEOS: TMSPM (1.0: 1.0), and (b) for increases of TMSPM molar ratios.

Fig. 2 shows the SEM cross section for films with MMA concentration of 0.25(a) and 2.0 (b). The SEM images show homogeneous films without any detectible defects. No phase separation is observed, which is consistent with the formation of cross-linked hybrid films between the organic and inorganic phases. The thickness of the samples, as determined by averaging several measurements, were 728 nm, 586 and 460 nm for the specimens with MMA molar ratio of 0.25, 1.0 and 2.0, respectively. The thickness of these hybrid films depend on the amount of PMMA contained in the sample, larger MMA concentration results in thinner films. This is likely due to the fact that the films with more MMA have an effective lower concentration of TMSPM, which results in solutions with lower viscosity and the concomitant thinner films.

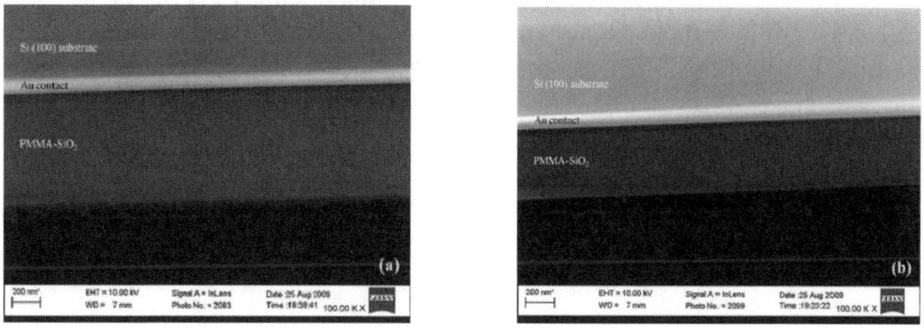

Fig. 2. Cross-section SEM micrographs of PMMA-SiO_2 films for samples with molar ratios of TEOS: TMSPM: MMA (a) 1.0: 0.25: 0.25 and (b) 1.0: 0.25: 2.0.

C-V analyses for MIM devices measured at several frequencies in the voltage range from -30 to 30 V for the three films are shown in Fig. 3. The MIM structure used for C-V analyses is shown in Fig 3(d). The different capacitance value observed is due mostly to the different film thicknesses.

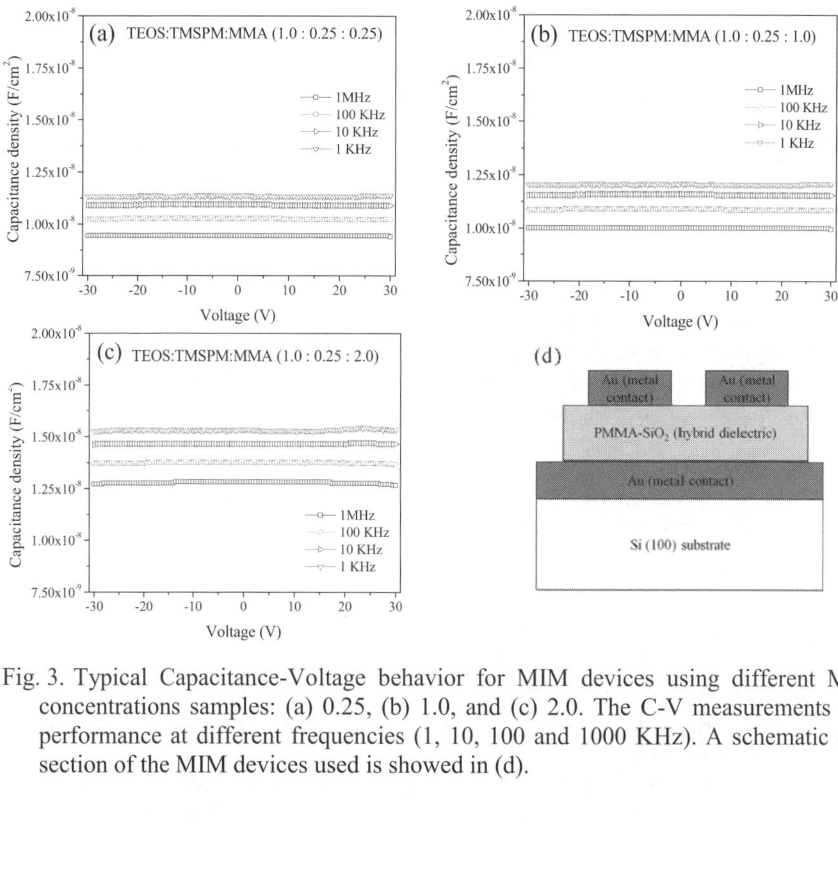

Fig. 3. Typical Capacitance-Voltage behavior for MIM devices using different MMA concentrations samples: (a) 0.25, (b) 1.0, and (c) 2.0. The C-V measurements were performance at different frequencies (1, 10, 100 and 1000 KHz). A schematic cross section of the MIM devices used is showed in (d).

The dielectric constant for each film as a function of frequency was calculated from these measurements according to Eq. 1 and the results are shown in Fig. 4. Typical dielectric constant values are 2.8-3.2 and 3.9 for PMMA and SiO_2, respectively. The dielectric constant at 1MHz for the PMMA-SiO_2 hybrid films was 7.8, 6.6, 6.7 for MMA: 0.25, 1.0 and 2.0 molar ratios, respectively. The higher dielectric constant is likely due to the presence of additional C-O-C bond from the TMSPM. These bonds are highly polarizable and has been reported to increase the dielectric constant [6]. The fact that films with lower MMA have higher dielectric constant seems to further corroborate this. Each TMSPM molecule contributes with 5 C-O-C bonds, whereas each MMA contributes with 2. Adding more MMA results in an effective "reduction" of the C-O-C bonds in the final hybrid films, with the concomitant reduced dielectric constant. This shows that the dielectric constant of PMMA-SiO_2 hybrid films can be tuned by adjusting the molar ratio of each component.

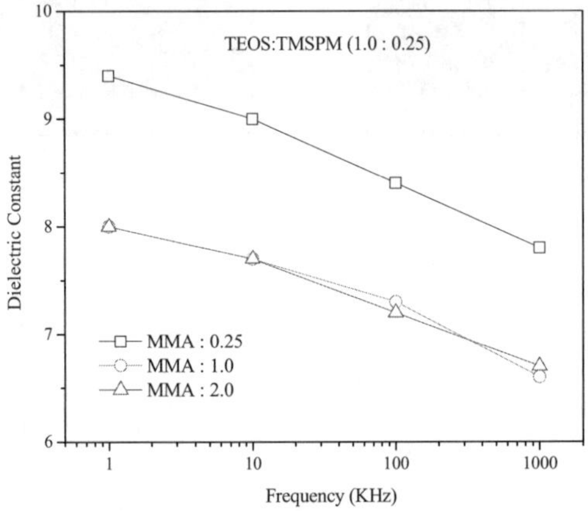

Fig. 4. Dielectric constant values at different frequencies for PMMA-SiO_2 hybrid material using different MMA molar ratios. The dielectric constant has a small variation (<10%) in the frequency range of 1 – 1000 KHz.

Conclusions

We studied the dielectric properties of PMMA-SiO_2 hybrid films. The dielectric properties of these films can be controlled by adjusting the molar ratio of each of its components. The films showed an increased dielectric constant (~7), which is likely related to the presence of additional C-O-C bonds from the TMSPM. The fact that these films were processed at temperatures <100°C indicates that this dielectric material can be used in flexible electronics.

Acknowledgements

This work was partially supported by Mexican Council of Science and Technology (CONACyT) under "becas mixtas" and sabatical program.

References

[1] C. Sanchez: J. Mater. Chem. Vol. 15 (2005) p. 3557.
[2] C. Sanchez, B. Julián, P. Belleville, M. Popall: J. Mater. Chem. Vol. 15 (2005) p. 3559.
[3] Sh. Al-Kandary, A.A.M. Ali, Z. Ahmad: J. Mat. Sci. Vol. 41 (2006) p. 2907.
[4] J.L. Almaral-Sánchez, E. Rubio E, A. Mendoza-Galván and R. Ramírez-Bon: J Phys. Chem. Solids Vol. 600 (2005) p. 1660.
[5] E. Rubio, J. Almaral, R. Ramirez-Bon, V. Castano, V. Rodriguez: Opt. Mat. 27 (2005) p. 1266.
[6] Moonkyong Na, Shi-Woo Rhee: Organic Electronic Vol. 7 (2006) p. 205-212.

Materials Science Forum Vol. 644 (2010) pp 29-32

doi:10.4028/www.scientific.net/MSF.644.29

Synthesis and Characterization of Magnetic Polyurethane Nanocomposite Foams

Alma Cámara-Hinojosa [a], Darío Bueno-Baqués [b], Oliverio S. Rodríguez-Fernández [c] and Ronald F. Ziolo [d]

Centro de Investigación en Química Aplicada, Blvd. Enrique Reyna No. 140, Saltillo, Coah. 25253, México

[a]acamarah@yahoo.com, [b]dbueno@ciqa.mx, [c]oliverio@ciqa.mx, [d]rziolo@cs.com

Keywords: polyurethane foams, magnetic, nanocomposite, synthesis, characterization.

Abstract. New magnetic polyurethane nanocomposite foams have been synthesized by the one-shot method. The opened-cell soft foams contain a dispersion of uniformly sized nanocrystalline iron oxide prepared *ex situ* by the high temperature thermal decomposition of iron acetylacetonate in phenyl ether in the presence of oleic acid. The magnetic particles were dispersed in polyol by sonication prior to the reaction of the latter with isocyanate to produce the magnetic foams. A 7 wt% loading of iron oxide yielded a soft, flexible foam with a room temperature magnetization of 3.5 emu/g at one Tesla. Higher loadings of iron oxide are possible without destruction of the soft open-cell polyurethane structure. Physicochemical characterization of the foams will be presented along with magnetic and mechanical properties. Potential applications include smart materials such as magnetic shape memory foams, adhesive-free metal to foam seals, inductively coupled thermal foams and applications for conformable foams having a magnetic component such as healthcare products.

Introduction

A polyurethane matrix containing magnetic nanoparticles offers the possibility of combining the properties of both components to formulate a composite material with unique properties. Magnetic nanoparticles could offer novel physicochemical properties arising from the core/shell structure and the particle interactions. Another side benefit of using magnetic nanoparticles as fillers that cannot be ignored is the nano-size scale in diameter [1]. Due to foams have obvious advantages, such as different thermal, magnetic and mechanical properties from the solid, including conformability, our objective was to synthesize new magnetic polyurethane nanocomposite foams, which have not been well studied and characterized. We chose to make opened-cell flexible polyurethane foams containing iron oxide nanoparticles. Nanocomposite foams were done by a one-step synthesis; meanwhile magnetic nanoparticles were synthesized by the high temperature decomposition of an organic precursor. We have characterized both nanoparticles and nanocomposites focusing on their magnetic and mechanical properties.

Experimental Section

Synthesis of Fe_3O_4 nanoparticles. $Fe(acac)_3$ (2 mmol), 1,2-tetradecanediol (10 mmol), oleic acid (6 mmol), and oleylamine (6 mmol) and phenyl ether (20 mL) were mixed and magnetically stirred under nitrogen flow. The solution was heated to 100°C and kept at this temperature for 45 min, followed by a second heating step at 200°C for 2 h. The solution was then refluxed for 1 h and subsequently cooled to room temperature by removing the heat source [2]. Under ambient conditions, methanol was added to the mixture and a black material was precipitated and separated magnetically. The black product was washed several times with methanol to remove any unreacted residue.

Synthesis of magnetic nanocomposites. An *ex-situ* one-shot polymerization method was used to synthesize the magnetic polyurethane nanocomposite foams with up to 7 wt% magnetite. Magnetic nanoparticles (0 to 7 wt%), were incorporated into a mixture of polyol, water and catalyst by sonication for 20 min. This was followed by the addition and mixing of isocyanate for 20 sec to make the magnetic polyurethane nanocomposite foams.
Characterization. XRD pattern of the magnetic nanoparticles were collected on a Siemens D-500 X-Ray Diffractometer using Cu Kα radiation (λ = 1.5405 Å). FTIR spectroscopy was obtained by using a Nicolet Magna 5500 Spectrometer. Samples for transmission electron microscopy (TEM) analysis were prepared by drying a toluene-nanoparticles dispersion on amorphous carbon-coated copper grid. The particles were imaged using an FEI Titan TEM at 300 kV. A Jeol Field Emission Scanning Electron Microscope was used to observe the structure of magnetic nanocomposites. Magnetic measurements were carried out on a Quantum Design PPMS using the susceptibility module to the AC susceptibility at 3 to 300 K and 50 Hz to 10 KHz and DC magnetization with fields up to 1 T at 3 and 300 K. A DMA Q800 Analyzer was used to study the viscoelastic behavior of nanocomposites and the effect of magnetic loading on their properties. Thermogravimetric analysis was carried out on a TGA Q500 (TA Instrument) at 10 °C/min under a N_2 atmosphere.

Results and Discussion

The thermal decomposition of the $Fe(acac)_3$ precursor in phenyl ether leads to the precipitation of particles with uniform size and shape (Fig. 1). The average diameter obtained from TEM data was 6.69 nm (σ = 0.23). XRD and TEM results confirmed an inverse spinel structure for the crystalline particles with lattice parameters similar to those of magnetite. IR spectral data along with TGA data indicated covalent bonding of the oleic acid to the particle surface [3].

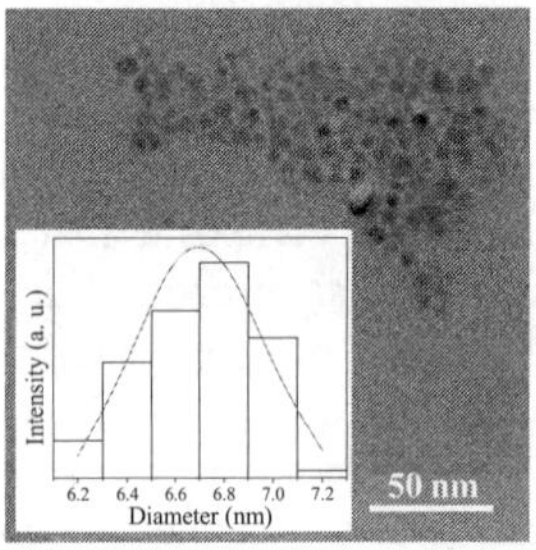

Fig. 1. TEM micrograph of oleic acid coated magnetite nanoparticles.

The magnetization data (Fig. 2) for the coated nanoparticles suggests superparamagnetic behavior at room temperature with no hysteresis and a saturation magnetization of 59 emu/g at 10 kOe. At 3 K, the particles showed a coercive field of 320 Oe, indicating ferromagnetic behavior for the nanoparticles with a saturation magnetization of 73 emu/g at 1 T. Lower saturation magnetization values to that reported for bulk magnetite (92 emu/g) could be due to the high surface area to volume ratio and therefore spin canting effects at the particle surface [4]. AC susceptibility measurements revealed that the particles followed the Vogel-Fulcher's relaxation model, $\tau = \tau_o \exp [E_o/k_B (T - T_o)]$. From the linear fit of $ln (\tau/\tau_o)$ vs $T_B\ ln (\tau/\tau_o)$ (Fig. 3), a value of T_o = 15.29 K indicates the presence of low interactions between particles; meanwhile, K_{eff} = 2.96x10^3 J/m^3 is close to that of bulk magnetite (10^4 – 10^5 J/m^3) [5].

The magnetic nanocomposite and pure polyurethane foams consist of open cells and spherical pores as shown in Fig. 4. Incorporation of magnetic particles provoked a decrease of about 38 % in pore size probably due to the oleic acid on the particle surfaces acting as a surfactant and thus decreasing superficial tension and/or the particles acting as a nucleation agent, provoking an increase in bubble generation, and, therefore, a decrease in pore size. FTIR results showed that the incorporation of magnetite into the polyurethane foam formulation does not affect its reaction outcome. On the other hand, the thermal stability of the nanocomposites was improved when magnetite nanoparticles were incorporated into the polyurethane formulation as seen by TGA. Presumably, this is due to the retarding effect of the magnetite nanoparticles on the movement of the polyurethane chains considering the inert nature of particles.

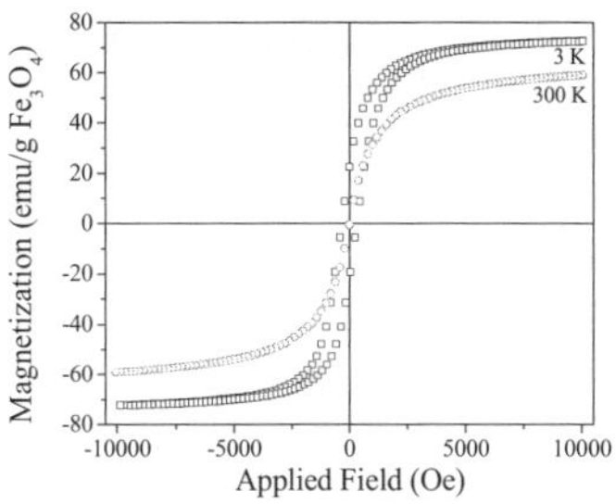

Fig. 2. Magnetization curve of oleic acid coated magnetite nanoparticles.

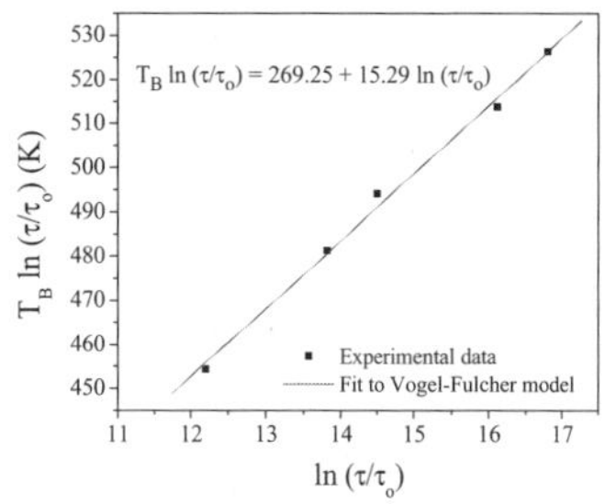

Fig. 3. Fit to the Vogel-Fulcher model of oleic acid coated magnetite nanoparticles.

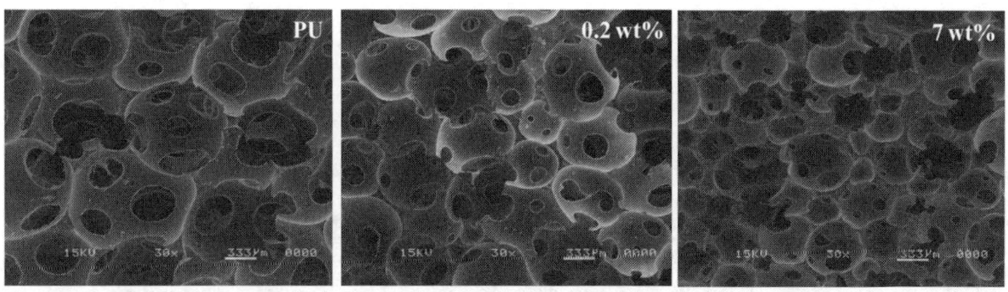

Fig. 4. Pure PU foam and oleic acid coated magnetite loaded PU nanocomposite foams.

The storage modulus of 0.2 wt% magnetite nanocomposite was 28 % higher than that of the pure polyurethane foam (Fig. 5), presumably due to the good dispersion and distribution of oleic acid coated magnetite nanoparticles in the polyurethane matrix and an increase in interfacial area. In contrast, at higher magnetite concentrations, the mechanical properties of the nanocomposites were smaller than pure polyurethane foam, due to the presence of particle clustering. Fig. 6 shows that T_g (glass transition temperature) increased from -29 °C for pure polyurethane foam up to -14.8 °C for 7 wt% magnetite nanocomposite. The increase of T_g could be attributed to an increase of adhesion between the polymer and magnetite nanoparticles because of the their size, which restricts segmental motion near the organic-inorganic interface. Also, T_g data indicate a rather substantial frictional force between the particle and the polymer as would be expected for a higher loading.

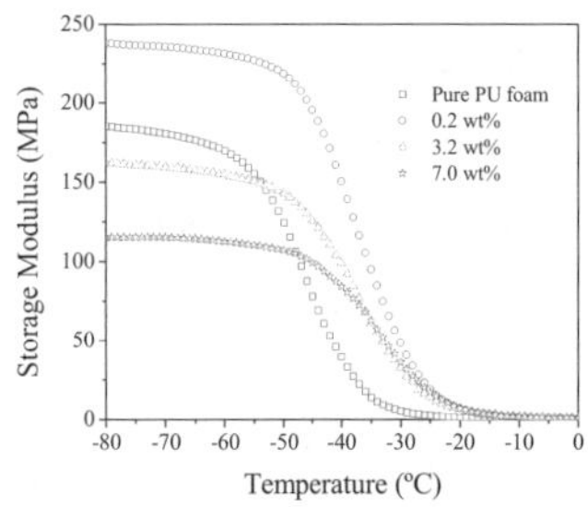

Fig. 5. Storage modulus curves of magnetic nanocomposites and pure PU foams.

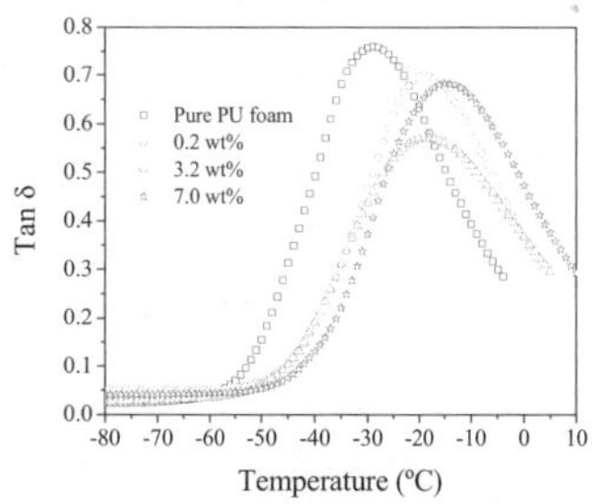

Fig. 6. Tan δ curves of magnetic nanocomposites and pure PU foams.

Magnetization curves of magnetic nanocomposites (Fig. 7) suggest that the nanocomposites behave superparamagnetically and ferromagnetically at room temperature and 3 K, respectively; following the same behavior of the magnetite nanoparticles. Saturation magnetization depends on magnetite concentration, increasing when the magnetite concentration is increased; the coercivity of the nanocomposite (300 Oe) was close to that of the nanoparticles. Regarding the magnetic relaxation, the AC susceptibility measurements revealed that the nanocomposites followed the Vogel-Fulcher relaxation model; the observed particle-particle interactions in the nanocomposites (T_o = 12.71 - 19.50 K) were similar to those observed in the nanoparticles (T_o = 15.29 K).

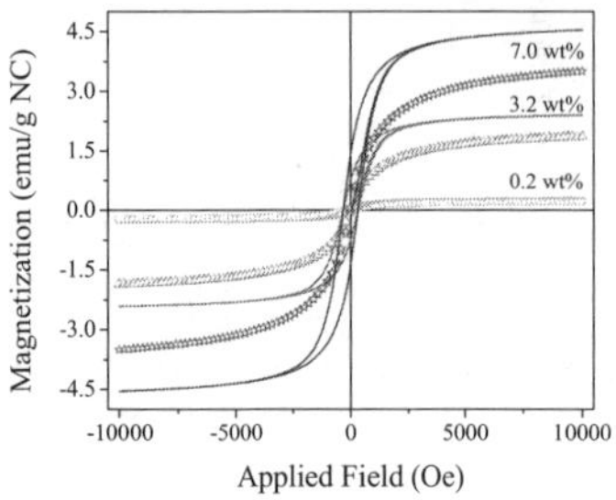

Fig. 7. Magnetization curves of magnetic polyurethane nanocomposite foams at 300 K (symbols) and 3 K (lines)

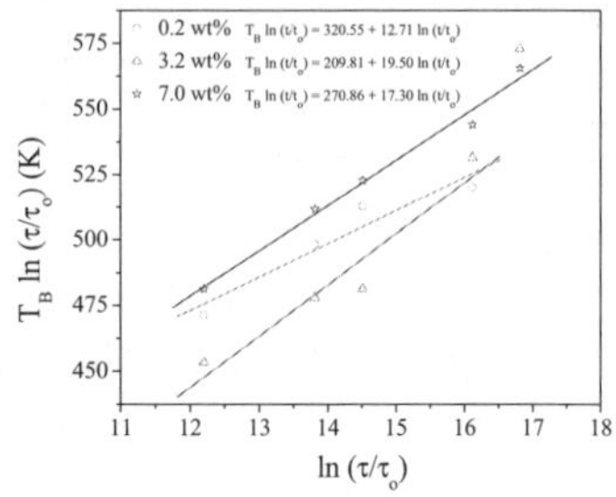

Fig. 8. Fit to the Vogel-Fulcher model of magnetic polyurethane nanocomposite foams.

Conclusions

Magnetite nanoparticles were successfully incorporated into polyurethane to produce magnetic nanocomposite foams with a higher thermal stability than the pristine foam; mechanical properties increased about 28% for a 0.2 wt% loading of magnetite. The magnetic nanocomposite properties depended linearly on the magnetite concentration with particle-particle interactions of the nanocomposites similar to those found in the magnetite powder.

Acknowledges

The authors would like to thank Dr. S. Moya (CICbiomagune, Spain) for TEM data and Dr. J. Tejada (University of Barcelona, Spain) and Gilberto Hurtado (CIQA, Mexico) for helping with magnetization measurements.

References

[1] Z. Guo, S. Park, H.T. Hahn, S. Wei, M. Moldovan, A.B. Karki and D.P. Young: J. Appl. Phys. Vol. 101 (2007), 511-513.

[2] M. Lattuada and T. A. Hatton: Langmuir Vol. 23 (2007), 2158-2168

[3] L. Zhang, R. He and H.C. Gu: Appl. Surf. Sc. Vol. 253 (2006), 2611-2617.

[4] A.G. Roca, M.P. Morales, K. O´Grady and C.J. Serna: Nanotech. Vol. 17 (2006), 2783-2788.

[5] R.M. Cornell and U. Schwertmann, *The Iron Oxides*, VCH, Weinheim-NY, 1996.

Materials Science Forum Vol. 644 (2010) pp 33-37
© (2010) Trans Tech Publications, Switzerland
doi:10.4028/www.scientific.net/MSF.644.33

Preparation of Electrospun Barium Titanate – Polyvinylidene Fluoride Piezoelectric Membranes

Verónica Corral-Flores[1,a], Jessica J. Pérez-Herrera[2,b], Enrique Torres-Moyc[3,c], Jorge Romero-García[1,d], Darío Bueno-Baqués[1,e], Ronald F. Ziolo[1,f]

[1]Research Center for Applied Chemistry, Blvd. Enrique Reina 140, Saltillo Coah. 25253, Mexico.

[2]Center for Research and Advanced Studies, Carr. Saltillo-Monterrey km 13.5, Coah. Mexico.

[3]Research Center for Advanced Materials, Miguel de Cervantes 120, Chihuahua, 31109, Mexico.

[a]vcorral@ciqa.mx, [b]jessy_jan13@hotmail.com, [c]enrique.torres@cimav.edu.mx, [d]jromero@ciqa.mx, [e]dbueno@mailaps.org, [f]rziolo@cs.com

Keywords: Piezoelectric, Electrospinning, Membranes, Nanocomposites.

Abstract. Piezoelectric fibrous membranes of barium titanate (BTO) – polyvinylidene fluoride (PVDF) nanocomposites were studied. BTO nanoparticles of about 40 nm were synthesized by the sol-gel method and mixed in PVDF–dimethyl formamide solutions at 0, 0.1, 1 and 3 wt.% relative to the polymer weight. The suspensions were electrospun using a horizontal set-up with an applied voltage of 15 kV. The samples were heat treated for 24 hours at 100ºC in air to increase the crystallinity of the polymer. The heat treatment induced a phase transformation from α to β phase in the pure polymer sample, while the nanocomposite membranes did not undergo such phase transformation. It was found that the addition of nanoparticles affected not only the morphology and diameter of the fibers, but also the content of beta phase of the polymer. In order to pursue the crystallization of β phase, additional samples were prepared by surface modification of the BTO nanoparticles and the addition of tetraisopentyl ammonium chloride. The best results were obtained with the last additive, which lead to the crystallization of only β phase and a homogeneous fibrous morphology. All these aspects were strongly correlated and consequently, governed the ferroelectric behavior of the samples.

Introduction

Polyvinylidene fluoride has been extensively and increasingly studied due to its ferroelectric, pyroelectric and piezoelectric properties [1,2]. It exhibits polymorphism depending on the chain conformations: TG^+TG^- in α and δ phases, all trans (TTT) planar zigzag in β phase, and $T_3G^+T_3G^-$ in γ and ϵ [3]. When ceramic nanoparticles are added to PVDF, ferroelectric nanocomposites with tailored properties can be obtained [4]. Specifically, the addition of barium titanate nanoparticles can enhance the piezoelectricity and electrical polarization due to its excellent piezoelectric properties [5]. BTO – PVDF composites have been attempted to tailor the dielectric properties of bulk composites [6,7], however, to the best of our knowledge, fibrous composite membranes showing piezoelectricity has not been reported yet.

Electrospinning technique offers the advantage of producing and poling the PVDF fibers in one single step, since the preparation involves high electric fields [8]. The positive effect of certain additives over the crystallization of beta phase has been studied in pure PVDF [9,10]. In the present work, we studied this effect over the crystalline beta phase on fibrous nanocomposites, which has been rarely attempted.

Experimental

Synthesis of barium titanate nanoparticles. $BaTiO_3$ was synthesized by sol-gel technique, starting from titanium butoxide and barium hydroxide monohydrate, in presence of acetic acid and stearic acid above 70 ºC. A clear gel was formed within minutes and it was overnight dried at 110°C in a stove. The dried gel was sintered at 700°C for 30 minutes.

Preparation of the piezoelectric composites. PVDF was dissolved in *N,N*-dimethyl formamide (DMF) in a volume ratio of 1:15. $BaTiO_3$ nanoparticles were dispersed in DMF at a weight ratio of 1:1 and added to the PVDF solution at proportions of 0.1, 1 and 3 wt.% with respect to the polymer weight. The mixture was sonicated for 1.5 hours and then stirred for 2 hours to improve the dispersion of the particles. These homogeneous suspensions were electrospun at a constant voltage of 15 kV using a horizontal set up. The distance between the electrodes was kept as 15 cm in all experiments, while the feed flux was 0.5 ml/hr. The samples were electrospun during 30 minutes to allow enough nanocomposite to be collected. Before the characterization, the membranes were vacuum-dried at room temperature for 24 hours. A portion of the samples was heat-treated at 100ºC for 24 hours.

Results and Discussion

An impurities-free perovskite phase of the barium titanate nanoparticles was confirmed by XRD. A refinement of the XRD data with PowderCell software revealed lattice constants a = 4.0037(1) Å and c = 4.0130(1) Å, thus the degree of tetragonality was calculated as 1.00232(6). The average particle size, determined also from the refinement was 40.7(4) nm.

The morphology of the as-spun membranes was investigated by SEM. The sample of PVDF alone showed straight fibers together with a certain number of beads, or small regions where the fiber gets thicker, as shown in Fig. 1(a). In contrast, all nanocomposites were conformed of thin fibers and a large amount of beads, as shown in Fig. 1(b) to (d). These results reflect the fact that the nanoparticles induced a drastic change in the morphology of the membranes, favoring the formation of beads. The average diameter of the fibers and the spheres is shown in Fig. 1(e), the error bars correspond to the standard deviation. Pure PVDF showed the highest values on both measurements, being nearly twice of those for the nanocomposites. As the amount of beads increased in the nanocomposites the diameter of the fibers decreased.

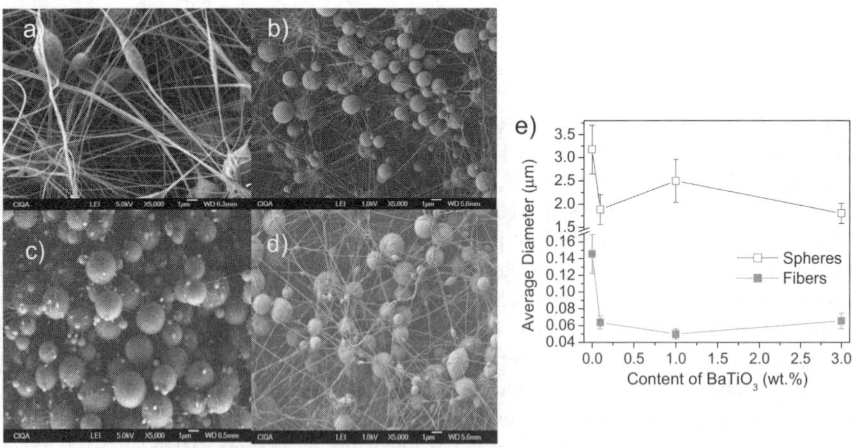

Fig. 1. SEM images of the as-spun $BaTiO_3$ - PVDF nano-composites with a) 0, b) 0.1, c) 1 and d) 3 wt.% of $BaTiO_3$ respect to PVDF; e) presents the average diameter of fibers and spheres.

Grazing incidence XRD revealed the presence of at least three crystalline phases on the as-spun samples, e.g. alfa, beta and gamma phases, regardless to the content of ceramic nanoparticles, as shown in Fig. 2 a). After heat treatment at 100°C for 24 hrs all samples showed higher crystallinity, as shown in figure 2 b). The heat treatment also induced a phase transformation from the α to the β phase. The sample without $BaTiO_3$ showed the most remarkable phase transformation, while the nanocomposites presented the beta phase at lesser extent. These results could be associated with the morphology of the samples. Previous studies revealed that the crystallization of beta phase requires not only heat treatment, but also unidirectional elongation, so the fibers seem to be more susceptible to undergo the phase transformation than the spheres due to its stressed nature. Thus, those samples with more bead concentration remain in its original crystalline phase after heat treatment, while those samples with more fibers than beads can crystallize in β phase.

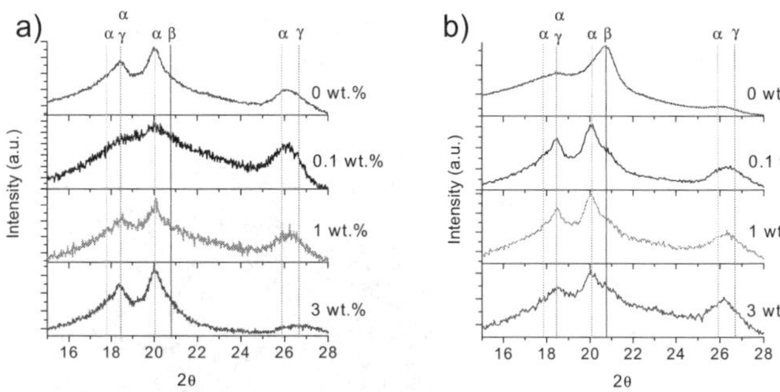

Fig. 2. GI-XRD patterns of the a) as-spun nano-composites and b) after heat treatment at 100°C for 24 hrs

FTIR in attenuated total reflection mode also reflected the presence of the three phases in all as-spun samples, as shown in Fig. 3(a). The absorption bands that correspond to α phase in the sample without ceramic nanoparticles are clearly diminished after heat treatment (Fig. 3(b)). These results are consistent with those obtained from GI-XRD. The fraction of β phase on the nanocomposites, estimated from FTIR-ATR spectra as proposed by Gregorio et al. [11], is shown in Fig. 3(c). The nanocomposites presented a tendency to diminish the fraction of beta phase after heat treatment, while the pure PVDF showed an increase of about 50%. This reflects the fact that the nanoparticles inhibit the crystallization of the beta phase, which is probably related to the sample morphology.

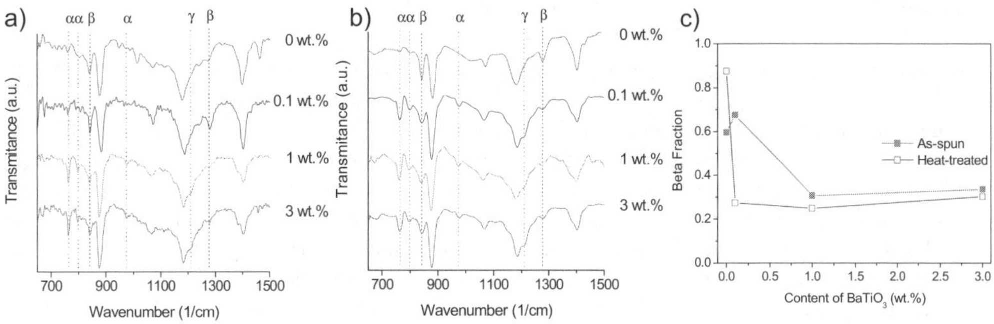

Fig. 3. FTIR-ATR spectra of the a) as-spun nanocomposites and b) after heat treatment at 100°C for 24 hrs. c) presents the fraction of beta phase estimated from FTIR-ATR results.

Enhancement of beta phase. Additional samples were prepared with the aim of enhancing the fraction of beta phase in the polymer. First, one sample was prepared with 3wt.% of tetraisopentyl ammonium chloride (TIPAC) with respect to PVDF. The preparation conditions were exactly the same than before. Another approach was the surface modification of the $BaTiO_3$ nanoparticles with two different silanes, e.g. metyl-trimethoxy-silane (m-silane) and aminopropyl-triethoxy-silane (a-silane). The nanoparticles were reacted with the silanes at 80ºC and then added to the PVDF solution following the same procedure described before. The amount of ceramic phase in these three samples was fixed to 1 wt.%, and the samples were characterized without heat treatment.

The fraction of beta phase was calculated from the FTIR-ATR spectra (shown in Fig. 4(a)). When a-silane and m-silane were used to modify the $BaTiO_3$ nanoparticles the fraction of beta phase was enhanced from 0.31 to 0.39 and 0.41, respectively. In contrast, when TIPAC was used as additive, the fraction of beta phase raised to 1, which indicates that the amount of alfa phase was completely suppressed. The mechanism responsible for such change in the crystalline phases is not well understood, but could be attributable to the ionic interactions of the polymer with the additive and the subsequent inhibition of the crystal growth.

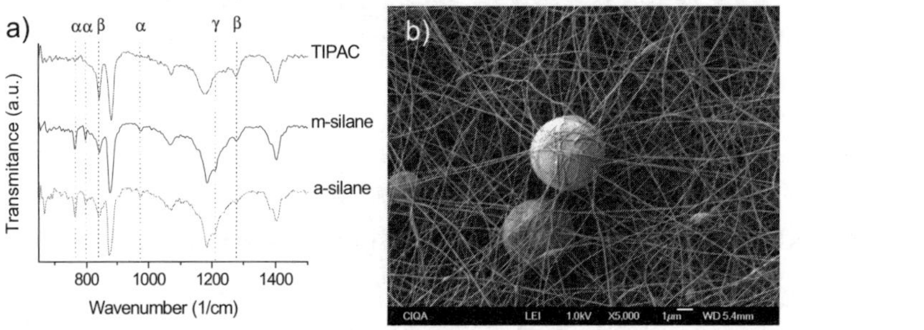

Fig. 4. a) FTIR-ATR spectra of the additional samples with 1 wt.% $BaTiO_3$ and b) SEM image of the as-spun sample with 1 wt.% $BaTiO_3$ and 3 wt.% TIPAC

The addition of TIPAC also induced a drastic change in the morphology of the sample, as shown in Fig. 4(b). The membrane is composed mainly by fibers, only few spheres were present. The average fiber diameter was 76(9) nm, while the average sphere diameter was 4.4(6) μm. Both sizes are bigger than the counterpart sample without additive. This morphology based on fibers could be responsible for the enhancement on the fraction of beta phase up to one, even without heat treatment.

Polarization measurements. The polarization measurements were done with a laboratory-made ferroelectric measurement station based on the virtual ground configuration. The polarization was obtained after a numerical integration of the charge derivative. The as-spun nanocomposites showed a quasi linear response, as shown in Fig. 5. In contrast, the sample with additive (1 wt.% $BaTiO_3$ and 3 wt.% TIPAC) showed a marked hysteresis, which is indicative of a ferroelectric behavior (see inset). Interestingly, the heat treated nanocomposites and those with surface modified BTO showed a similar linear response, even for the sample without ceramic nanoparticles, which showed a high fraction of beta phase. These results suggest that the addition of $BaTiO_3$ nanoparticles did not improve the ferroelectric properties of the membranes, while the addition of TIPAC dramatically enhanced the ferroelectric behavior. This could be attributed to the presence of the additive itself, or could be directly related to the morphology of the membrane.

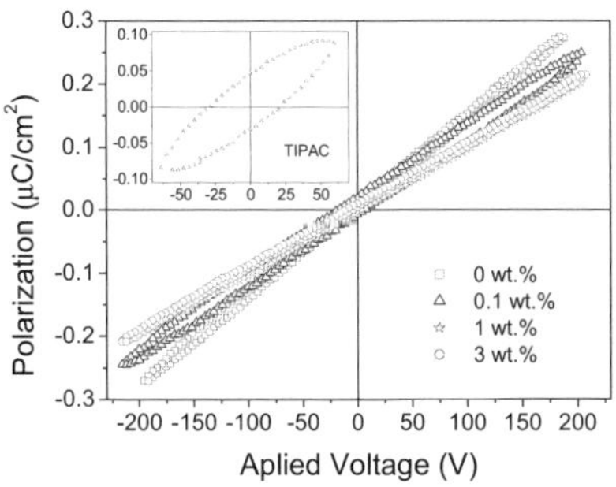

Fig. 5. Electrical polarization of the as-spun nanocomposites. The inset shows the hysteresis curve of the sample with 1 wt.% $BaTiO_3$ and 3 wt.% TIPAC.

Summary

Piezoelectric fibrous membranes of barium titanate (BTO) – polyvinylidene fluoride (PVDF) nanocomposites were studied. Heat treatment at 100ºC for 24 hrs induced a phase transformation from α to β phase only in the pure polymer sample. The addition of nanoparticles affected the morphology, fiber diameter, and the content of beta phase of the polymer. The addition of tetraisopentyl ammonium chloride in a nanocomposite with 1wt.% BTO lead to the crystallization of pure β phase and a homogeneous fibrous morphology. The electrical polarization measurements showed a quasi linear response for both the as-spun and the heat treated membranes. In contrast, the sample with 3 wt.% TIPAC showed a marked ferroelectric hysteresis.

References

[1] M. Benz, W.B. Euler: J. Appl. Polym. Sci. 89 (2003), p. 1093

[2] M.C. Branciforti, V. Sencadas, S. Lanceros-Mendez, R. Gregorio Jr.: J. Polym. Sci. B: Pol. Phys. 45 (2007), p. 2793

[3] A. Salimi, A.A. Yousefi: J. Polym. Sci. B: Pol. Phys. 42 (2004), p.3487

[4] H.C. Pant, M.K. Patra, Aditya Verma, S.R. Vadera, N. Kumar: Acta Mater. 54 (2006), p. 3163

[5] H. Basantakumar Sharma, H.N.K. Sarma, A. Mansingh: J. Mat. Sci. 34 (1999), p.1385

[6] R. Sharma, I.P. Singh, A.K. Tripathi, P.K.C. Pillai: J. Mat. Sci. 29 (1994), p. 995

[7] Z.M. Dang, L.Z. Fan, Y. Shen, C.W. Nan: Mat. Sci. Eng. B103 (2003), p.140

[8] Z. Zhao, J. Li, X. Yuan, X. Li, Y. Zhang, J. Sheng: J. Appl. Polym. Sci. 97 (2005), p. 466

[9] M. Nasir, H. Matsumoto, M. Minagawa, A. Tanioka, T. Danno, H. Horibe: Polymer Journal 39, 7 (2007), p. 670

[10] W.A. Yee, M. Kotaki, Y. Liu, X. Lu: Polymer, 48 (2007), p. 512

[11] R. Gregorio Jr., M. Cestari: J. Polym. Sci. B: Pol. Phys. 32, 5 (1994), p. 859

Materials Science Forum Vol. 644 (2010) pp 39-42
© (2010) Trans Tech Publications, Switzerland
doi:10.4028/www.scientific.net/MSF.644.39

Preparation and properties of $CoFe_2O_4$ synthesized by the modified citrate-gel method

E.M. Múzquiz-Ramos[1,2,a], D.A. Cortés-Hernández[1,b], O.A. Herrera-Romero[1,c], J.C. Escobedo Bocardo[1,d]

[1]CINVESTAV IPN-Unidad Saltillo, Carr. Saltillo-Mty, Km 13, A.P. 663, C.P. 25000, Saltillo, Coah. México

[2]Facultad de Ciencias Químicas, Universidad Autónoma de Coahuila, Blvd. V. Carranza y José Cárdenas, C.P. 25280, Saltillo, Coah. México.

[a]emuzquiz@uadec.edu.mx, [b]dora.cortes@cinvestav.edu.mx, [c]oziel_romero@hotmail.com, [d]jose.escobedo@cinvestav.edu.mx

Keywords: Cobalt ferrite; Modified citrate-gel method; Nanoparticles; Magnetic properties.

Abstract. In this work, the synthesis of $CoFe_2O_4$ via sol-gel auto-combustion method, using iron nitrate, cobalt nitrate and citric acid, with subsequent heat treatment in air was studied. The effects of the molar ratio of the metal nitrates to citric acid and the heat treatment temperatures on the magnetic properties have been investigated. The X-ray diffraction patterns showed peaks consistent with cubic spinel-type structure. The average crystallite sizes were determined from the (311) peak of the diffraction pattern using Scherrer equation. Particle sizes in the range of 18-44 nm were obtained. The crystallite size increases with annealing temperature. Magnetic properties, such as saturation magnetization (Ms), remanent magnetization (Mr) and coercivity field (Hc) were measured at room temperature using a vibrating sample magnetometer. Saturation magnetization was found to increase with particle size, whereas coercivity was found to reduce exponentially as the particle size was increased, apart from the Hc of the sample treated at 300 °C. The present work shows that magnetic properties vary over a wide range by changing the synthesis conditions.

Introduction

Hyperthermia is a therapeutic procedure used to raise the temperature in a region of the body affected by cancer. It is administered together with other cancer treatment modalities (multimodal oncological strategies). The required hyperthermic temperature can be achieved by various methods. There is a wide range of hyperthermia applications which uses radiofrequency, microwave, ultrasound or infra-red (superficial and whole body hyperthermia). Each of them has its limitations and many of them can be used either externally or internally. The most effective system is that in which the ferrite particles for non-invasive heating are used. Ferrite particles under an alternating magnetic field are expected to be useful as thermoseed in hyperthermic cancer treatment since they can be targeted and confined to the cancer site without damaging normal cells [1,2]. These magnetic particles have to be nanometric and superparamagnetic or soft ferrimagnetic.

$CoFe_2O_4$ is a typical ferrimagnetic oxide with spinel structure. For many years, it has been the subject of research activities because of fundamental matters as well as its potential applications. Ferrimagnetic $CoFe_2O_4$ exhibits excellent magnetic properties [3], good specific loss power (SLP) in alternating magnetic field [4] and non-cytotoxicity [1]. Also it is known that most of the physical properties of these ferrites depend strongly on the shape and size of the nanosized particles [5-7]. These materials presented considerable differences in magnetic properties when compared with their bulk materials [8].

Several methods have been reported to synthesize cobalt ferrite nanoparticles, these include polyol method [9], sol gel [10,12], combustion reaction [11], solid state reaction, citrate precursor [12] and co-precipitation [13]. The modified citrate-gel method has shown to be useful for obtaining cobalt ferrites that heat more efficiently tumors, without damaging normal cells, in a low magnetic field, however, there is a controversy about the synthesis conditions [3,14].

In order to obtain nanosized Co-ferrite which can be used in hyperthermia therapy, in this work, nanometric $CoFe_2O_4$ was prepared using a sol-gel auto-combustion method. The aim was to study the influence of the heat treating temperature on the magnetic properties and crystallite size using a molar ratio 1:3 of the metal nitrates to citric acid and pH 7.

Experimental

The nanoparticles of cobalt ferrite were synthesized by auto-combustion reaction using analytical grade iron nitrate ($Fe(NO_3)_3 \cdot 9H_2O$, 99.99%, Aldrich), cobalt nitrate ($Co(NO_3)_2 \cdot 6H_2O$, 98%, Aldrich), citric acid ($C_6H_8O_7$, 99.5%, Aldrich) and ammonium hydroxide (NH_4OH, 28-30%, Aldrich). The metal nitrates were disolved together in a minimum amount of double distilled water to get a clear solution, the molar ratio of Co:Fe was 1:2. An aqueous solution of citric acid was mixed with the metal nitrates solution, the molar ratio betweeen total metal ions and citric acid was controlled at 1:1, 1:2, and 1:3. The mixture containing the salts, deionized water, and citric acid was stirred to obtain a homogeneous solution and then was heated to 80 °C at a heating rate of 5 °C/min using a hot plate magnetic stirrer. The pH of the solution was adjusted to 7 by dropping ammonia. The sol was formed during stirring and the temperature of the hot plate was incresed up to 150 °C at constant stirring, the water in the solution was evaporated and a viscous gel was formed. When finally all water molecules were removed from the mixture, the viscous gel began frothing. After few minutes, the gel automatically ignited and burnt with glowing flints. The decomposition reaction would not stop before the whole citrate complex was consumed. The auto-ignition was completed within a minute, yielding the brown-colored ashes termed as a precursor.

The samples were heat treated at temperatures between 300 and 1200 °C for 1 h in air to form cobalt ferrites nanoparticles. The particles were analyzed by using a vibrating-sample magnetometer (VSM, Lakeshore 7307) and X-ray difractometer (XRD, X'Pert, Phillips, Holland).

Results and discussion

Fig. 1 shows the XRD pattern of cobalt ferrite prepared in a molar ratio of 1:3. Only peaks of pure-phase cubic spinel structure of $CoFe_2O_4$ were detected (JCPDS 22-1086) [15]. As the temperature was increased the intensity of the characteristic peaks was also increased. No other phases, such as hematite, were observed by XRD. The increase of the annealing temperature leaded to a higher crystallization without changing the nature of the cobalt ferrite.

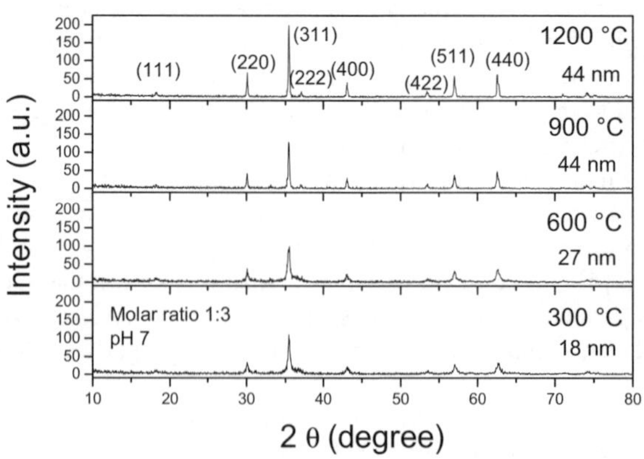

Fig. 1. XRD patterns of the samples heat treated at 300, 600, 900 and 1200 °C for 1 hour.

The particle size was determined by the Scherrer equation using the most intense (311) peak [16]. The average particle size is shown in Table 1. As expected, the particle size increases with annealing temperature. The hysteresis loops of samples heat treated at 900 °C, with different molar ratio between total metals ions and citric acid are shown in Fig. 2; the coercive field decreases with increasing the molar ratio and the magnetization values increase up to a maximum of 68 emu/g. In all cases, the values are lower than the reported for bulk $CoFe_2O_4$ (80 emu/g) [18]. Due to these results and taking into account that a low coercivity is required for hypertermia treatment [19], the selected molar ratio was 1:3. Hysteresis loops at room temperature (Fig.3) show a drastic change in shape from the first sample annealed at 300 °C which is ferrimagnetic, to that observed for the sample treated at 1200 °C which shows a softer magnetic behavior.

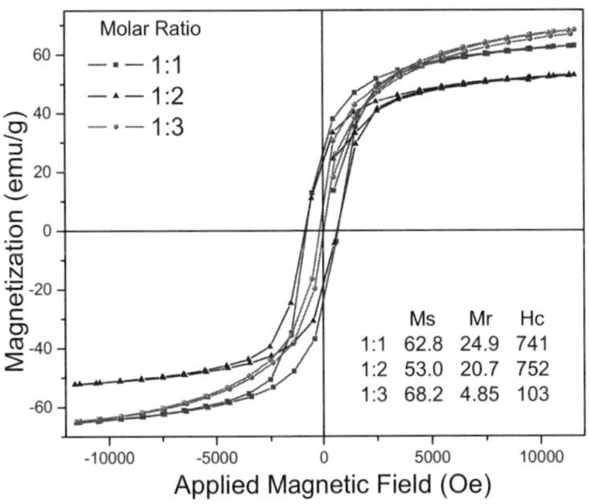

Fig. 2. Hysteresis loops of the samples heat treated at 900 °C.

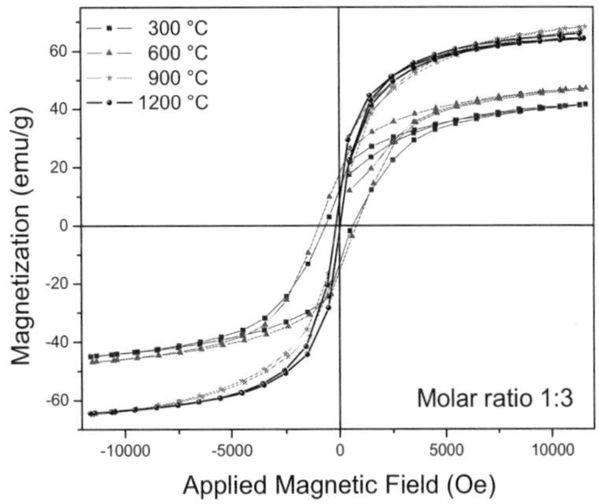

Fig. 3. Hysteresis loops of the samples heat treated at different temperatures (molar ratio 1:3).

The saturation magnetization (Ms), coercive field (Hc) and remanent magnetization (Mr), along with the average particle size of the samples synthesized in a molar ratio of 1:3, are summarized in Table I.

Table I.
Average particle size and magnetic properties at room temperature of $CoFe_2O_4$ obtained in a molar ratio of 1:3.

Heat treatment condition	Average particle size [nm]*	Ms [emu/g]	Mr [emu/g]	Hc [Oe]
300 °C	18	41	12	643
600 °C	27	47	16	891
900 °C	44	68	5	103
1200 °C	44	85	4	99

* determined from Scherrer formula

An increase in particle size takes place by increasing temperature, the lattice defects and strains decrease, this may cause coalescence of crystallites that results in increasing the average of the nanoparticles size [17]. As it is well known [5-7], the differences in the magnetic properties are associated with changes in particle size, since magnetic moments increase with particle size. When samples are treated at 600ºC or at higher temperatures, magnetization increases with temperature, while the coercive field and remanent magnetization decrease.

Conclusions

Nanocrystalline magnetic particles of $CoFe_2O_4$ with different particle size ranging from about 18 to 44 nm were obtained by the modified citrate-gel method. By changing molar ratio and annealing temperature, nanoparticles with different magnetization values and particle sizes can be prepared.

The size and the magnetic properties of these Co-ferrite particles indicate that these can be successfully used as thermoseeds for hyperthermia under an alternating magnetic field.

Aknowledgements

E. M. Múzquiz Ramos thanks CONACyT-Mexico for the financial support (scholarship 267018)

References

[1] D.H. Kim, S.H. Lee, K.N. Kim, K.M. Kim, I.B. Shim and Y.K. Lee: J. Magn. Magn. Mater. 293 (2005) 287–292.

[2] S. Somiya, S. Aldinger, N. Claussen, R.M. Uchino, K.Koumoto, M. Kaneno in: *Handbook of Advanced Ceramics Vol II. Processing and their applications*. Edited by Elsevier Academic Press. (2003) p 394.

[3] R. Sato Turtelli, G.V. Duong, W. Nunes, R. Grössinger, M. Knobel: J. Magn. Magn. Mater. 320 (2008) 339–342.

[4] D.H. Kim, S.H. Lee, K.N. Kim, K.M. Kim, I.B. Shim and Y.K. Lee: J. Magn. Magn. Mater. 293 (2005) 320–327.

[5] E.K. Mooney, J.A. Nelson, M.J. Wagner: Chem. Mater. 16 (2004) 3155.

[6] D.R. Cornejo, A. Medina-Boudri, H.R. Bertorello, J. Matutes-Aquino: J. Magn. Magn. Mater. 242 (2002) 194.

[7] M. Rajendran, R.C. Pulla, A.K. Bhattacharya, D. Das, S.N. Chintalapudi, C.K. Majumdar: J. Magn. Magn. Mater. 232 (2001)71.

[8] L.A. García, V.A. Torres, J.A. Matutes, O.E. Ayala: J. Alloy. Compd. 369 (2004) 148–151.

[9] G. Baldi, D. Bonacchi, C. Innocenti, G. Lorenzi, C. Sangregorio: J. Magn. Magn. Mater. 311 (2007) 10–16.

[10] S.M. Montemayor, L.A. García, J.R. Torres, O.S. Rodríguez: J Sol-Gel Sci. Techn. (2007) 42:181–186.

[11] A. Franco, E.C. de Oliveira, M.A. Novak, P.R. Wells Jr.: J. Magn. Magn. Mater. 308 (2007) 198–202.

[12] P.C. Rajath, R. Sekhar, D. Banerjee, M. Raama , K.G. Suresh, A.K. Nigam: J. Alloy Compd. 453 (2008) 298–303.

[13] R. Betancourt, O. Ayala, L.A. García, O. Rodríguez, J. Matutes, G. Ramos, H. Yee: J. Magn. Magn. Mater. 294 (2005) 33–36.

[14] B.G. Toksha, S.E. Shirsath, S.M. Patange, K.M. Jadhav: Solid State Commun. 147 (2008) 479–483.

[15] JCPDS International Centre for Diffraction Data. Mineral Powder Diffraction File. Data Book. USA (1990).

[16] B.D. Cullity, S.R. Stock in: *Elements of X-Ray Diffraction*. Third Edition. Edited by Prentice Hall, New Jersey, (2001) p. 167-171.

[17] T.P. Raming, A.J.A. Winnubst, C.M. Van Kats, P. Philipse: J. Colloid Interf. Sci. 249 (2002) 346.

[18] J. Gwang Lee, H. Min Lee, C. Sung Kim, Y. Jei Oh: J. Appl. Phys. 84 (5) (1998) 2801.

[19] T.Atsumia, B.Jeyadevanb, Y.Satob, K.Tohjib: J. Magn. Magn. Mater 310 (2007) 2841–2843.

Materials Science Forum Vol. 644 (2010) pp 43-46
© (2010) Trans Tech Publications, Switzerland
doi:10.4028/www.scientific.net/MSF.644.43

Alumina-Copper Composites with High Fracture Toughness and Low Electrical Resistance

E. Rocha-Rangel[1,a] and J. G. Miranda-Hernández[2,c]

[1]Departamento de Mecatrónica, Universidad Politécnica de Victoria,
Luis Caballero No. 1200, Col. Del Maestro, 87070, Cd. Victoria, Tamaulipas, México
[2]Departamento de Materiales, Universidad Autónoma Metropolitana,
Av. San Pablo No. 180, Col. Reynosa-Tamaulipas, 02200, México, D. F.

[a]erochar@upv.edu.mx, [b]jgmh@correo.azc.uam.mx

Keywords: Alumina-Copper Composites, Toughness, Electrical Resistance

Abstract. Through an intense mixing process of Al_2O_3 powder with different copper contents, Al_2O_3-Cu composites were fabricated by sintered at 1300°C during 1h, where the likely liquid sintering mechanism, lead to obtain composites with relative densities greater than 95%. Scanning electron microscopy was used to observe the resulting microstructures, which indicated that these composites were mostly formed by a fine and homogeneous Al_2O_3-ceramic matrix with immerse nano-metallic copper particles. The behavior of both fracture toughness and electrical resistance of the composites is directly dependent with the copper content in the matrix. As the copper contents increased, the composites exhibited high values of fracture toughness, whereas, their electrical resistance is reduced considerably.

Introduction

Alumina (Al_2O_3) exhibits favorable physical and chemical properties such as; high strength, hardness, elastic modulus and excellent resistance to thermal and chemical environments. However, its applications are somewhat limited because of poor toughness and inferior thermal shock resistance [1]. On the other hand, most structural Al_2O_3-based ceramics present poor electrical conductivity. It has been reported that the incorporation of some amounts of fine-size metal particles into an Al_2O_3 matrix, as in the case of Al_2O_3/Ni [2], Al_2O_3/W [3] and Al_2O_3/Cu [4] composites, can significantly improve both the fracture toughness and electrical conductivity.

Accordingly, the requirement to produce ceramics with both high toughness and electrical properties has given rise to the necessity to fabricate new ceramic systems reinforced with metallic particles and produced through the development of new methods, among them is the mechanical alloying. Although, this method was developed since the 60's, is only until few years ago that it has started to be used importantly in the processing of new advanced materials, because with its application it becomes possible to obtain composite materials with unusual microstructures and consequently properties [5, 6].

The purpose of this study is to fabricate Al_2O_3-Cu composite materials with different copper contents, through the combination of mechanical alloying and liquid sintering, and to determine the effect of copper on the fracture toughness and electrical resistance of the fabricated composites.

Experimental Procedure

Through an intense mixing process of Al_2O_3 powder (1μm. 99.9%, Aldrich) with different copper metal (5μm. 99.9%, Aldrich) contents, (1, 5, 10, 20 and 30 wt. %), several Al_2O_3-Cu composites were fabricated. The mixture was prepared in dry using a ceramic ball mill (15 cm diameter) with ZrO_2 media (0.5 cm diameter), at a fixed rotation speed of 300 rpm during 24 h. The ball-to-powder volume ratio was 25:1. With the milled powder mixture, cylindrical samples having 2 cm diameter and 0.3 cm height were fabricated by uniaxial pressing using 270 MPa. Then the pressed samples were sintered for 1h at 1300°C in an electrical furnace in argon atmosphere. During sintered, the samples were heated and cooled at 10°C/min rates. The fracture toughness was estimated by the fracture indentation method [7], whereas the electrical resistance of the composites was measured by the four points probe technique [8]. Finally, the microstructure was observed by scanning electron microscopy (SEM). It is important to mention that for each one of the studied systems, five samples were made and the results reported here are the averages of the measurements made in each set of samples.

Results and Discussion

Properties

Fig. 1 shows the fracture toughness (K_{IC}) and electrical resistance changes as a function of the copper content in the composites. From Fig. 1 it can be appreciated that the sample's fracture toughness increases with the increasing copper content in the composite. It is clear that the incorporation of ductile copper inside the hard alumina matrix increase the alumina toughness. It was reported [9] that the most likely toughening mechanism is the crack bridging due to the presence of a homogeneous ductile metal network in the material's microstructure, as can be observed in the Fig. 2. Fig. 1 also displays the electrical resistance results as a function of the copper content in the composites. It can be observed that the electrical resistance reduces with the increments of copper content. This can be explained as follows; the results obtained in the composites indicate that the quantity of electrical current under different voltages applied follows a lineal behavior. This allows the determination of the electrical resistance through Ohm's Law, applying a lineal fitting procedure to the experimental data. From the current-voltage experimental data it is estimated the electrical resistance. Comparing these results with the electrical conductivity of the elements constituting the composite it is found that: The alumina has an electrical resistivity $>10^{14}$ Ω-cm [10]. From the data it can be calculated that its electrical conductivity is about $<10^{-14}$ (1/ Ω-cm), this means that it is very difficult for the electricity to pass through it. If the resistance is now computed considering the sample's dimensions, the resistance is $3.4802X10^{13}$ Ω. This means that the incorporation of copper in the ceramic matrix diminishes its electrical resistance in almost 50% for the sample with 30 wt. % Cu.

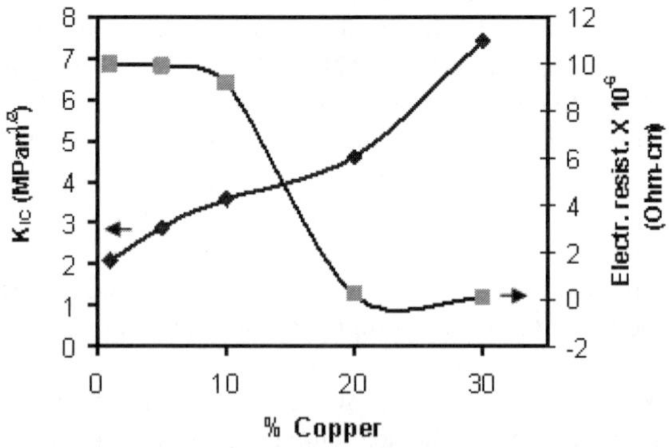

Fig. 1. Fracture toughness and electrical resistance changes as a function of the copper content in the composites. Samples sintered at 1300°C for 1h.

Microstructure

Fig. 2a reveals that for Al_2O_3-1% Cu composite has a homogeneous microstructure with respect to the grain size. Also, Fig, 2a shows a fine microstructure, with a very low quantity of the present copper in the sample (clear phase). It can be noticed also that the porosity is present uniformly.

In the case of Al_2O_3-5% Cu composite, it can be observed that the same characteristics of the microstructure are shown in Fig. 2b, that is, a clear gray phase that corresponds to the matrix grains is apparent, a white and brilliant phase due to the metal and dark hollows that correspond to porosity. It becomes significant that, the copper metallic phase is more noticeable, found forming elongated thin shapes in the grain surroundings in the alumina matrix that are shaping partially the metallic net as a result of the copper increase in the composition, which is taking up some of the porosity spaces in the porosity, thus diminishing it.

Fig. 2c shows that in Al_2O_3-10% Cu composite, the metallic network is still partial and isolated, although is more visible due to the greater metal content in the composition. Another appreciably observation is the growth of both grains and pores size.

Fig. 2d shows that in Al_2O_3-20% Cu composite the grain in the microstructure is now bigger, the pores are larger but their number has diminished.

The previous discussed aspects are relevant in the development of the microstructure of the composites obtained; however the presence of a higher metal content in the Al_2O_3-30% Cu composite makes the diffusion in the compact more feasible as shown in Fig. 2e. This give rise to improve density as the porosity reduced, which is correlated with both the fracture toughness and electrical behavior. The formation of the Cu network brings changes in density and open porosity. It is important to consider that densification is a diffusion process and diffusion is a mass transfer phenomena. So if there is copper in the system and this metal is a very good heat conductor, by analogy this means that copper can transport mass easily. So mass transfer is enhanced with the copper presence, by a sintered liquid mechanism, and for this reason the grain growth in the samples with the increments in the copper content is then explained.

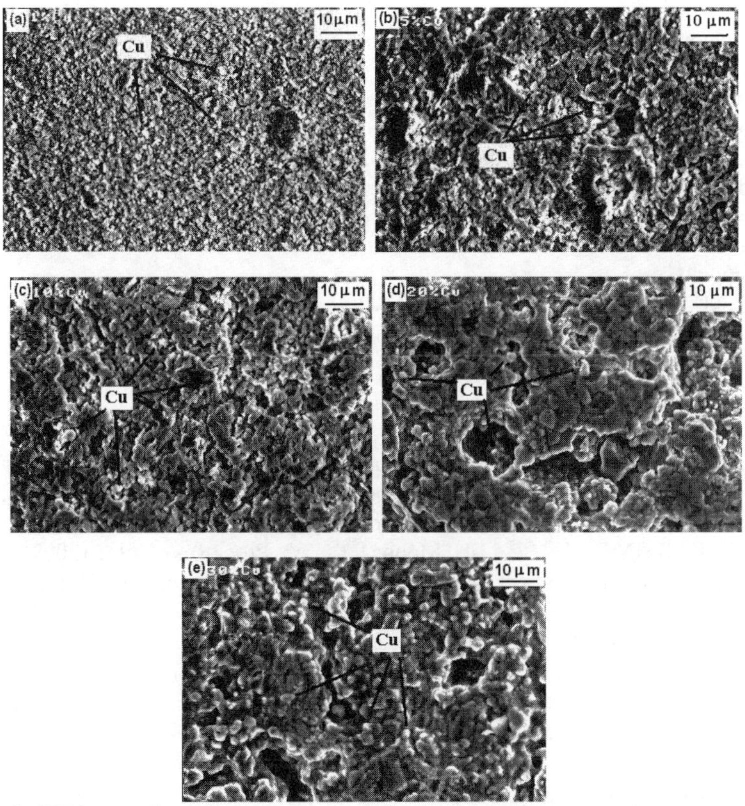

Fig. 2. SEM secondary electron image analysis of the composites microstructural as a function of the copper content. Samples sintered at 1300 °C for 1 h. a) Al_2O_3-1 wt. % Cu, b) Al_2O_3-5 wt. % Cu, c) Al_2O_3-10 wt. % Cu, d) Al_2O_3-20 wt. % Cu and e) Al_2O_3-30 wt. % Cu.

With the help of energy dispersive of X-ray (EDX) in the SEM, it can be detected that the microstructure is formed by an Al_2O_3-ceramic matrix (dark phase) with fine Cu metallic networks (clear and brightness phase) which the results are shown in Table 1. Because the sintered temperature was 1300 °C, copper became liquid because it melts at 1083 °C, therefore, the operating sintering mechanisms took place with the presence of a liquid phase. The liquid helps to obtain dense composites because it goes into the material's bulk filling the sites previously occupied by pores, thus forming an interpenetrating network with the ceramic matrix. At this respect, it is important to mention than several authors [11, 12,] have documented good wetting of Al_2O_3 ceramic by liquid copper.

Table 1. Quantitative analysis of Cu network for each sample.

Theoretical wt. % Cu content in the sample	Actual wt. % Cu content in the sample
1	0.8
5	4.7
10	9.6
20	19.3
30	29.1

Summary

Al_2O_3-Cu composites with different copper contents were produced successfully by the combination of mechanical alloying and liquid sintered. The composites' microstructures are constituted by an Al_2O_3-ceramic matrix interconnected with a fine and homogeneous copper metallic network. The incorporation of a ductile metal inside a hard ceramic matrix improves its fracture toughness. The probable toughening mechanism is the crack bridging due to the presence of homogeneous ductile metal networks in the composites' microstructure. On the other hand, copper and the network that it forms also help to reduce the electrical resistance of the composites.

Acknowledgment

Authors would like to thank to the various foundation of promotion to the scientific and technological investigation from CONACyT-Tamaulipas Government, by the support offered for the accomplishment of the research work.

References

[1] J. K. Wessel: The Handbook of Advanced Materials, John Wiley & Sons, New York. 2004.
[2] T. Sekino, T. Nakajima, S. Ueda and K. Niihara: J. Am. Ceram. Soc., Vol. 80 (1997), p. 1139.
[3] T. Sekino and K. Niihara: Nanostructal Materials, 2ed., New York (1995).
[4] S.J. Ko, K.H. Min, Y.D. Kim and I-H. Moon:,J. Ceram. Proc. Res., Vol. 3 (2002), p. 192.
[5] P. H. Shingu: Thermodynamic principles of metastable phase formation, First Int. Conf. on Proc. Mats. For Properties; The Minerals, Metals & Materials Society, (1993), p. 1275.
[6] Ighodaro O.L. - Okoli O.I.:. Int. J. Appl. Ceram. Technol. (2008), pp. 313-323.
[7] A.G. Evans and E.A. Charles: J. Am. Ceram. Soc., Vol. 59 (1976), p. 371.
[8] Information on: http://www.kilowattclassroom.com/Archive/GndTestArticle.pdf. (2006)
[9] M. M. El-Sayed Seleman, J. Mater. Sci. Technol., 24 (2008) 723-728.
[10] J.F. Shackelford and W. Alexander: Materials Science and Engineering Handbook. CRS Press, Boca Raton Florida (2001).
[11] J. M. Howe: International Materials Reviews, Vol. 38 (1993), p. 233.
[12] J. G. Li: J. Amer. Ceram. Soc., Vol. 75 (1992), p. 3118.

Materials Science Forum Vol. 644 (2010) pp 47-50
© (2010) Trans Tech Publications, Switzerland
doi:10.4028/www.scientific.net/MSF.644.47

Obtaining NiHCF Nanoparticles Using a Reverse Micellar System

Alberto Gutiérrez Becerra, Tanya Vega Venegas, Maximiliano Bárcena Soto, Norberto Casillas and José Inés Escalante[a]

University of Guadalajara, Department of Chemistry, Boul. Marcelino García Barragán 1451, Guadalajara, Jalisco, 44430 – México

[a]escalant@hotmail.com

Keywords: Nickel ferrocyanide nanoparticles, reverse microemulsion, modified surfactant.

Abstract. The synthesis of metallic nanoparticles using a reverse microemulsion medium brings controlled size and shape. In this study, reverse micelar system was employed for the synthesis of nickel hexacyanoferrate nanoparticles, NiHCF, by direct encapsulation in the water pools of reverse micelles formed in a pseudoternary system that consist of a mixture of 95 %wt. cetyltrimetyl ammonium bromide (CTAB) and 5 %wt. of a novel modified surfactant cetyltrimetyl ammonium ferrocyanate ($CTA[Fe(CN_6)]^{4-}$, CTAFe)/*n*-butanol/Hexane/Water. These NiHCF nanoparticles were produced with particle size ranging with 20 – 60 nm.

Introduction

The synthesis of metal nanoparticles by reverse micelles is viable and attractive because not only does it produce nanoparticles that have a narrow size distribution but also the particle sizes can be easily controlled by varying the microemulsion (ME) composition [1]. This synthesis may be conducted in two modes: (i) the simple addition, where the reducing or precipitating reagent is directly added to the ME containing the other reactant, and a multiple ME-method, where two or more MEs, each containing a reactant, are mixed together; (ii) the synthesis by replace the counterion of the surfactant, and then only the addition of a salt to this reverse ME-media [2].

This work deals with the replace the counterion of the surfactant method. The Br^- ions of the cationic surfactant CTAB are replaced by $[Fe(CN)_6]^{4-}$ ions following a direct metathesis reaction in aqueous phase, thus forming a new modified surfactant named hereafter as (CTAFe) [3]. By doing so one is able to incorporate the complex counterion without having to add an additional salt. However, it is possible to produce different metal complex compounds, e. g. NiHCF, CoHCF, CuHCF etc. by simple adding different metal-cations to the aqueous phase.

Experimental Section

Materials. All the reactants used in this report were analytical grade. Cetyl trimethyl ammonium bromide (CTAB) was purchased from Sigma-Aldrich (99%), ferrocyanide salt ($K_4[Fe(CN)_6]\cdot 3H_2O$) from J.T. Baker (99%), *n*-Hexane (C_6H_{14}) from Caledon Laboratorios LDT (98%), *n*-butanol (C_4H_9OH) from Productos Químicos Monterrey (99%), and double distilled water from Selectropura S.A. de C.V. (σ = 1.5 – 3 μS/cm). CTAFe was prepared by a direct metathesis reaction in aqueous phase and characterized by FTIR (see figure 1a, where were compared FTIR spectra of pure CTAB and CTAFe). One see that both spectra appear to have the same structure, just there are two peaks characteristics of CTAFe: one due to the Fe – C vibration (at 545 cm^{-1}) and the other one due to the C≡N stretching (at 2000-2100 cm^{-1}) [3,4].

***Electrical Conductivity* (σ).** The samples were prepared by weighting appropriates amounts of the surfactants, $W_{Surf} = W_{CTAB}+W_{CTAFe}$, and n-butanol ($W_{Surf} : W_{n\text{-butanol}}$ :: 1), and then an adequate amount n-hexane was added. All the samples were placed into a test tube at 25 °C, kept at constant stirring and titrated with double distilled water with 0.1mL additions. The complete incorporation of each water addition into the mixture allowed performing σ-measurements directly with a conductimeter (Thermo Orion 4-Stars Series, USA). Repeated experiments in all the samples were performed with less than 2% variation in σ.

Synthesis of NiHCF. In order to verify the formation of nanoparticles of NiHCF in this system, one proof was carried out (see point in figure 1). In 4 g of a system with dilution line, H, of 0.4 (where H is defined as the amount of n-hexane, W_{Hexane}, to a fixed amount of surfactant and cosurfactant, i.e. $H = W_{\text{Surf}}+W_{\text{n-butanol}}/W_{\text{Surf}}+W_{\text{n-butanol}}+W_{\text{Hexane}}$), was added under stirring 0.4g of solution 0.005M $NiCl_2$. The ME formed after the addition of the solution was stable for several days. The X-ray diffraction (XRD) patterns were collected on a diffractometer (Siemens, mod. D-500). This equipment has a graphite monocromator using CuKα radiation at a wavelength 0.154 nm. Using an HRTEM JEOL 2010 performed TEM micrographs of the nanoparticles of NiHCF prepared inside the ME droplets. The dispersion was placed on a carbon-coated copper grid. The grid was dried at room temperature for 10 min prior to the analysis. Samples exposed to the electron beam for a prolonged duration were extensible damaged.

Results and Discussion

Phase behavior. The pseudo-ternary phase diagram obtained for the CTAB + CTAFe/n-butanol/n-hexane/water at 25 °C is shown in Figure 1. A simple titration technique was used to construct this diagram. The surfactants are insoluble in n-hexane and to reach the w/o ME region it is necessary to add a small amount of water ($\approx$ 2%wt). Further addition of water ($\approx$ 40%wt.) made the system pass through this region up to the appearance of permanent turbidity, indicating that the system reached the maximum amount of water that can be solubilized. The w/o ME region was determined as being the one where permanent transparency was observed.

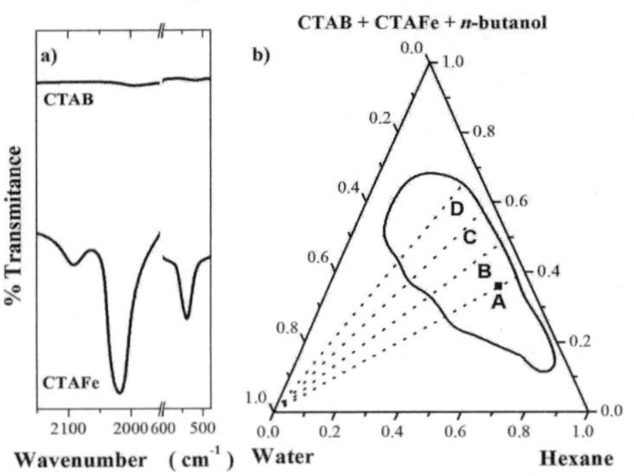

Figure 1. a) FTIR of CTAB and CTAFe; **b)** Pseudo-ternary phase diagram obtained for the CTAB + CTAFe/n-butanol/n-hexane/water at 25 °C. The dilution lines, H, that were used for electrical conductivity measurement are also shown: A) 0.4, B) 0.5, C) 0.6 and D) 0.7. The filled square in this figure is the composition at which NiHCN nanoparticle was synthesized.

ME structure. MEs of differing compositions were prepared in the one-phase region: four sets were prepared with increasing aqueous phase weight fractions (0.40, 0.50, 0.60 and 0.70; while the surfactant and hydrocarbon weight fractions were adjusted accordingly). In order to determine the compositional domains of w/o MEs, electrical conductivity was measured after each water addition of the titration procedure. Figure 2a shows σ versus the weight fraction of water (W_w) as function of the dilution line. Notice that for small amounts of added water low values of σ are observed. As soon as the amount of water in the system reached the percolation threshold, W_w^c (see figure 2b), the specific conductivity increased almost linearly with the amount of water added. The percolation threshold (W_w^c) is defined as the peak in a plot of the rate of change in the log value of σ with W_w of the reverse micelle solution. This type of behavior is typical of w/o structures for which, as the weight fraction increases or as the aqueous volume fraction increases, the micelles start clustering and their ions are rapidly exchanged with each other; therefore σ sharply increases [5]. The percolative conduction phenomenon observed in some w/o MEs can be explained in terms of "sticky droplet collisions", as proposed by Fletcher and Robinson [6]. Thus, above the percolation threshold, the "sticky collisions" between droplets would result in the formation of water micro-channels through which ions could move producing a sharp increase in σ.

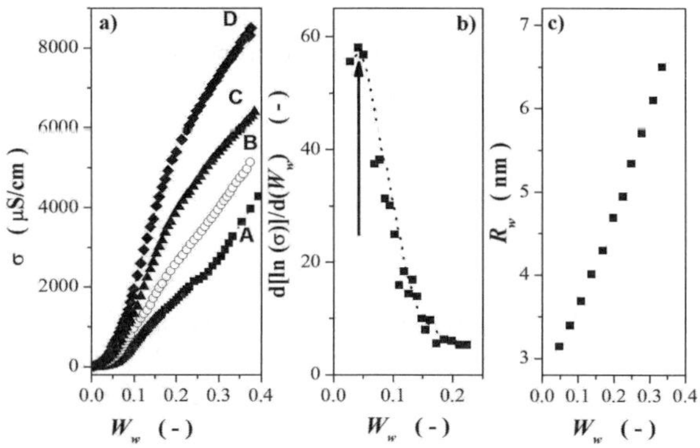

Figure 2. **a)** Electrical conductivity (σ) vs. W_w to different H values (■: 0.4, ●: 0.5, ▲: 0.6, ♦: 0.7) at 25 °C; **b)** Rate of change in the log value of σ with W_w for H of 0.4. Arrow points the percolation threshold (W_w^c); **c)** Tendency of the droplets core radio, R_w, as function of W_w.

By using σ-data, together with the Effective Medium Theories, EMT, we obtained the average radius of the droplets core, R_w [7]. One observes an increase of R_w with increasing W_w; this is an expected result, since small droplets can swell or coalesce forming larger ones (figure 2c). This variation is expected because the presence of butanol as a cosurfactant in the interface reduces the electrostatic repulsion between the $-N^+(CH_3)_3$ groups in the polar heads of the surfactants, thus the distance between them becomes smaller [8]. This phenomenon also favors the flexibility of the interface. Therefore, as soon as more butanol molecules are incorporated into the interface, the sizes of the droplets become smaller. The radii obtained in all the relations are of the same order of magnitude to the ones reported in the literature and ranges from 3 to 9 nm [9,10]. To satisfy this rationale, for a constant [CTAFe] in the ternary mixture, finer droplet dispersion with increasing number should arise with decreasing dispersant (water) volume, i.e., with decreasing W_w.

NiHCF Nanoparticles. The formation of NiHCF-nanoparticles in this system is reached when the solution of $NiCl_2$ is added to the w/o ME. These NiHCF nanoparticles synthesized inside the droplets of the w/o ME are shown in the figure 3. The reaction between the $[Fe(CN)_6]^{4-}$ ions from the CTAFe and the Ni^{2+} ions from the aqueous solution produces a yellowish brown colored ME without precipitation, suggesting that this reaction is sufficiently facile to induce the growth of the NiHCF. The inset of the figure 3 shows a XRD pattern of the NiHCF particles obtained into the droplets of the w/o ME. The well-defined peaks seen in this inset indicate that NiHCF is crystalline. The diffraction data have been compared with the standard data file [11]. It is clear that the sizes of the particles are larger than the expected by the size of the droplets (≈ 8nm diameter). This behavior has been reported for other HCF's nanoparticles synthesized in w/o ME with AOT [12,13]. The principal reason of this behavior is the low concentration of reactants inside the droplets, *e.g.* at a 0.005 M of $NiCl_2$, only about six Ni^{2+} ions are present in each 8 nm diameter water droplet, so the growth process must involve thousand of droplet exchange, as well as an increment in the droplet size. Another reason could be just an aggregation process during the sampling.

Conclusions

A novel modified cationic surfactant (CTAFe) was used successfully to form nanoparticles of NiHCF, which was obtained by replacing the Br^- ion by the coordination compound ion ferrocyanide. The phase diagram for the pseudo-ternary CTAB+CTAFe+*n*-butanol/*n*-hexane/water system shows a wide inverse ME region that extends from low water concentration (2%wt.) to medium water concentrations (40%wt.). The assumption that the system is formed by spherical aggregates, which may form clusters after the concentration percolation threshold, is fulfilled for MEs with $H \geq 0.4$ and σ-data matched correctly the EMT. The average radius of the droplets core, R_w, increases with increasing W_W; and results to be of the same order of magnitude to the ones reported in the literature and ranges from 3 to 9 nm [9,10]. The formation of NiHCF-nanoparticles inside the droplets of the w/o ME was performed and their sizes are larger than the expected by the droplet size (≈ 8nm diameter). This behavior has been reported for other HCF's nanoparticles synthesized in w/o ME with AOT [12,13]. Finally, the main advantage of the proposed system is the possibility of synthesizes several MHCF by varying the transition metal.

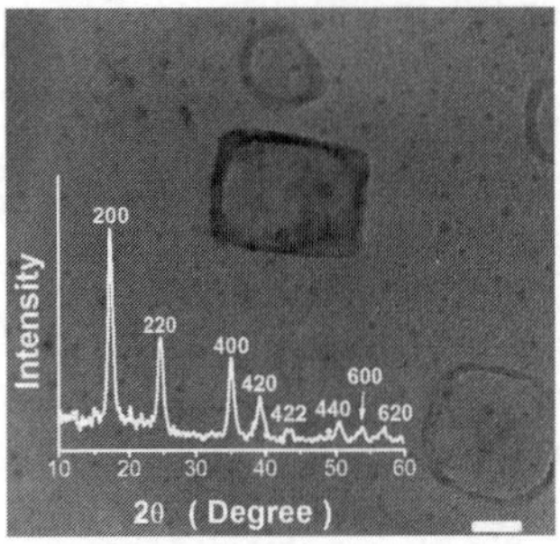

Figure 3. TEM micrograph of NiHCF nanoparticles synthesized in w/o ME. Inset: A typical XRD Electron pattern of the same sample. The Bar represents 20 nm.

Acknowledgements

This work was supported partially by the Mexican National Council of Science and Technology (CONACyT) and the German Academic Exchange Service (DAAD) within the framework of the international cooperation project (PROALMEX grant I0110/127/08).

References

[1] K. Osseo-Asare: Microemulsion-mediated synthesis of nanosize oxide materials, Ed. by P. Kumar and L. Mittal, Handbook of Microemulsion Science and Technology (Marcel Dekker, Inc., New York, 1999) p. 549-603.

[2] L. Motte: J. Mater. Sci. Vol. 31 (1996), p. 38.

[3] A. Gutierrez-Becerra, T. Vega-Venegas, M. Barcena-Soto, J.I. Escalante, P. Malo de Molina, S. Prévost, M. Gradzielski: Langmuir (2009) submitted.

[4] G. Le Caër, G. Vigneron, J.P. Renault, S. Pommeret: Chem. Phys. Lett. Vol. 426 (2006), p. 71.

[5] B. Lagourette, J. Peyrelasse, C. Boned, M. Clausse: Nature Vol. 281 (1979), p. 60.

[6] P.D.I. Fletcher, B.H. Robinson: Ber. Bunsen-Ges. Phys. Chem. Vol 85 (1981), p. 863.

[7] G. Gu, W. Wang, H. Yan: J. Coll. Int. Sci. Vol. 178 (1996), p. 358.

[8] R. M. Pashley, P. M. McGuiggen, B. W. Ninham: J. Phys. Chem. Vol. 90 (1986), p. 5845.

[9] S. Paul, S. Bisal, S.P. Moulik: J. Phys. Chem. Vol. 96 (1992), p. 896.

[10] S.P. Moulik, W.M. Aylward, R. Palepu: Canadian J. Chem. Vol. 79 (2001), p. 1.

[11] Index to the Powder Diffraction File: ASTM Technical Publication, 48-M_2, 1963.

[12] S. Vaucher, J. Fielden, M. Li, D. Dujardin, S. Mann: Nano letters Vol. 2 (2002), p. 225.

[13] S.P. Moulik, G.C. De, A.K. Panda, B.B. Bhowmik, A.R. Das: Langmuir Vol. 15 (1999), p. 8361.

Materials Science Forum Vol. 644 (2010) pp 51-55

doi:10.4028/www.scientific.net/MSF.644.51

Iron oxide nanoparticles obtained from a Fe(II) - chitosan polymer film.

J.F. Luna-Martínez[1, a], E. Reyes-Melo[1,2, b], V. González-González[1,2, c], A. Torres-Castro[1,2, d], C. Guerrero-Salazar[1,2, e], S. Sepúlveda-Guzmán[1,2, f].

[1]FIME-UANL, Av. Universidad s/n Cd. Universitaria San Nicolás de los Garza, N.L., 66450, México
[2]CIIDIT-PIIT, Nueva Autopista al Aeropuerto Internacional Monterrey Km 10, Apodaca, N.L., México.
[a]juan.fco.luna.mtz@gmail.com, [b]mreyes@fime.uanl.mx, [c]virgilio.gonzalezgnz@uanl.edu.mx, [d]atorres@mail.uanl.mx, [e]carlos.guerreros@uanl.mx, [f]selene.sepulvedagz@uanl.edu.mx

Keywords: Polymer film, chitosan, Superparamagnetic behavior, Magnetic nanoparticles.

Abstract. In this work, iron oxide nanoparticles (~5 nm) embedded in a chitosan polymer film, were synthesized. In order to obtain this nanostructured material, firstly a homogenous film of Fe(II)-chitosan was prepared. The resulting composite film has a thickness of ~140μm. Iron oxide nanoparticles were in-situ synthesized by treating the composite film with H_2O_2 under alkaline conditions. The morphological analysis by Transmission Electron Microscopy (TEM) shows the nanoparticles were embedded and stabilized in chitosan polymer film. The magnetic behavior was studied by magnetization measurements. The magnetization curves at room temperature showed that iron oxide nanoparticles have a superparamagnetic behavior.

Introduction

The relevance of nanomaterials is due to the fact that; particles with size ranging from 1nm to 10nm exhibit novel electronic, optical, magnetic and chemical properties [1]. Iron oxide nanoparticles are of particular interest due to their fundamental properties associated to their multivalent oxide state and polymorphism [2]. Several technological applications have been reported for these nanoparticles, for example, magnetic data storage [3], catalysis [4], magnetic fluids [5], medical diagnosis [6], controlled drug delivery [7], and magnetic-induced cancer therapy [8]. Iron oxide magnetic nanoparticles have been synthesized by different methods such as sol-gel [9], chemical coprecipitation [10] and microemulsion [11]. The tendency of these magnetic nanoparticles to form clusters during the synthesis process is usually decreased by the precipitation of the nanoparticles in an inter matrix. Several materials have been used as matrix in the preparation of iron oxide nanocomposites such as silica [12], zeolites [13], and polymers [14]. In this work, we paid attention to the synthesis and characterization of iron oxide nanoparticles embedded in a chitosan film acting as polymer matrix. The most remarkable characteristic of chitosan is its biocompatibility, which makes of this material adequate for several biomedical applications [15]. chitosan acts as stabilizing-agent, avoiding the formation of iron oxide clusters. The morphological analysis and structural characterization were done by (TEM) and (XRD) respectively. In addition, we studied the magnetic behavior by magnetization measurements and field cooling and zero field cooling analysis.

Experimental

Medium molecular weight chitosan with a deacetylation degree of 75-85% was supplied by Aldrich The ferrous chloride tetrahydrate ($FeCl_2{\cdot}4H_2O$, 99.9 %) was supplied by Fermont. Formic acid (88%) was purchased from Jalmek and used without purification. In a typical process 0.2 g of chitosan were dissolved in 20 ml of concentrated formic acid and stirred for 30 minutes. Then 15 ml of 0.028M of $FeCl_2{\cdot}4H_2O$ in formic acid were added to the chitosan solution. The mixture was kept at 40°C under stirring for 3 h. The solution was poured in a glass Petri dish and the solvent was evaporated at 40°C and at atmospheric pressure, for 20 hours obtaining a homogenous film of Fe(II)-chitosan. The thickness of the resulting film obtained using a mitutoyo micrometer was ~140 μm. The Fe(II)-chitosan film was placed in a 1.8M solution of NaOH at 40°C observing a color change of the film from brown to black color. Then 20ml, of 3% H_2O_2 solution were added to the black film and after 15 minutes the film was removed from the mixture. The color of the film turned back to brownish color. The film was washing with distilled water and ethanol in order to eliminate the residual salts such as NaOH and NaCl. The brownish film was characterized by X-ray diffraction using a Bruker Advanced Solution D8 diffractometer. Transmission electron microscopy analysis was done in a Jeol 2010operating at 200 KV of voltage accelerating. The magnetic properties were measured at both room temperature (VSM, Model 7300 Lake Shore) and low temperature, using a Quantum Design MPMS5 SQUID magnetometer.

Results and discussion

The composite polymer film based on Fe(II) and chitosan was analyzed by X-ray diffraction and the resulting diffraction pattern is shown in Figure 1. The $FeCl_2{\cdot}4H_2O$ diffraction peaks (Figure 1c) were undetected in the diffraction pattern of the Fe(II)-chitosan polymer film (Figure 1b) suggesting a complete dissolution of the salt in chitosan, and the formation of a complex coordination between them [16]. Also, it can be seen three broad peaks that could be associated to some ordering of the chitosan during the film formation in presence of Fe(II) ions but more experiments should be done in order to corroborate that.

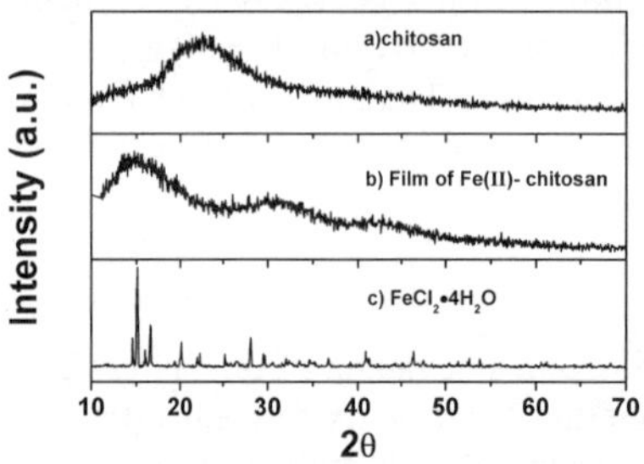

Fig.1. X-ray diffraction patterns for (a) as received chitosan, (b) Fe(II)-chitosan polymer film and (c) as received $FeCl_2{\cdot}4H_2O$ salt.

After treating the Fe(II)-chitosan polymer film with NaOH and H_2O_2, the change in the color of the film suggests the formation of nanocomposite of Iron oxide-chitosan polymer film [1]. This nanocomposite was analyzed by TEM, and Figure 2 shows the TEM micrograph. Both, size and nearly spherical morphology of the iron oxide nanoparticles embedded in the chitosan polymer matrix are noticed. The average particle size was of 5nm measured from the TEM images, and the histogram is shown in Figure 3. The TEM images also show the well dispersed nanoparticles in the polymer film, indicating efficient particle size stabilization by chitosan matrix. This result could be possible due to strong interaction between amino groups in chitosan chains and Fe(II) or Fe(III) ions on iron oxide nanoparticle surface, and also it could be due to the presence of hydroxyl groups on nanoparticle surface [17]. In our group we have been synthesized magnetic nanoparticles with chitosan as polymer matrix and we have observed that particle size distribution is dependant of the synthesis process [18]. In this case, smaller particle size and narrow size distribution were obtained due to the previous preparation of the Fe(II)-chitosan polymer film.

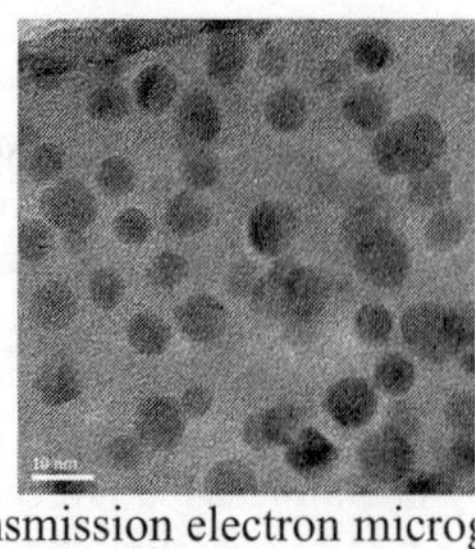

Fig. 2. Transmission electron micrograph of the magnetic nanocomposite.

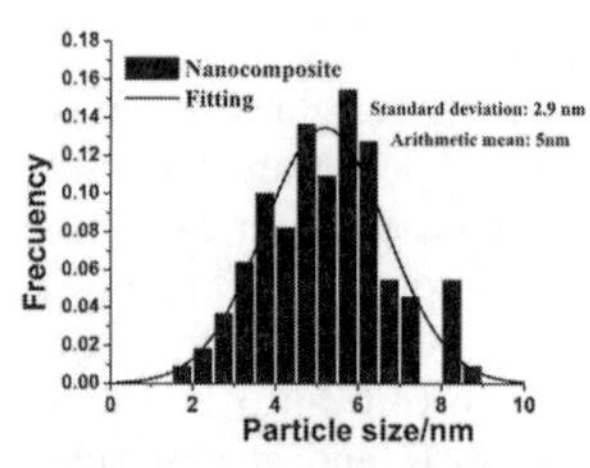

Fig. 3. Particle size histogram obtained from TEM micrographs.

The Iron oxide-chitosan polymer film was analyzed by XRD to know the crystalline structure of the iron oxide nanoparticles. Figure 4, shows the diffraction pattern of as synthesized nanocomposite and it shows broad diffraction peaks due to the small particle size. It was difficult to identify the crystalline structure of the iron oxide nanoparticles in the nanocoposite film from the resulting diffraction pattern. The diffraction peaks can be associated to both maghemite and magnetite phases by comparing them to the standard patterns JCPDS No. 39-1346 and JCPDS No. 19-0629 respectively. Nevertheless some peaks agree very well with maghemite crystalline structure and are indexed in the diffraction pattern (Figure 4). We believe that during the synthesis process after the addition of sodium hydroxide the precipitation of iron oxide occurs and when the H_2O_2 solution was added the remaining Fe(II) ions were oxidized to Fe(III), and for this reason the iron oxide phase in the nanocomposite film was identified as maghemite. The synthesis of maghemite nanoparticles using a similar method has been widely studied [1]. In addition, it can be observed two strong peaks at 20° and 21°, which are associated with iron oxide hydroxides species (α-FeOOH) usually present during the synthesis of iron oxides by precipitation method [14].

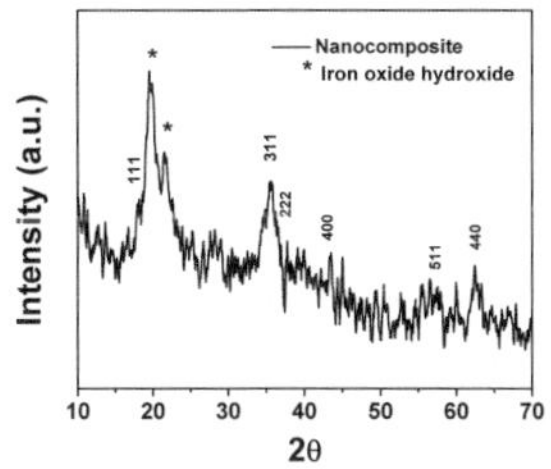

Fig. 4. X-ray diffraction pattern of the resulting nanocomposite.

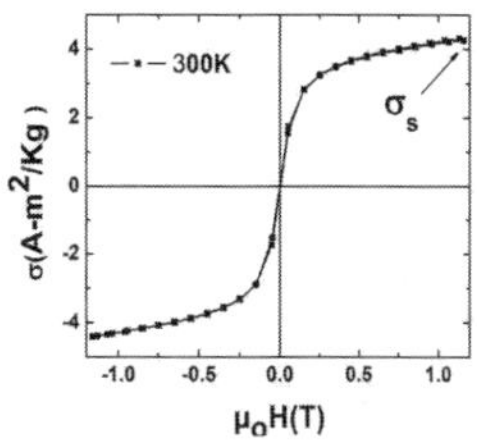

Fig. 5. Room temperature magnetization curve for Iron oxide-chitosan polymer nanocomposite at $1.16\mu_0 H(T)$, where μ_0 is the permeability of the vacuum.

Figure 5 shows the magnetization curve at room temperature (300K) of the Iron oxide-chitosan nanocoposite film. The saturation magnetization value (σ_s) is 4.25 A-m^2/kg whereas both remanent magnetization and coercive field (H_C) are not discernible. These results, absence of coercive field and magnetic remanence, are characteristics of a magnetic nanocomposite with superparamagnetic behavior [19].

On the other hand, Figure 6 shows the magnetization curve obtained at 10K. Under these conditions both coercive field and remanent magnetization values are different to zero. From the magnetization curve it can be seen that the saturation magnetization value is 6.75 A-m^2/kg, coercive field (Hc) is of 7.7 μ_0H(mT) and the remanence value (σ_r) is 1.0 A-m^2/ kg; (inset figure 6).

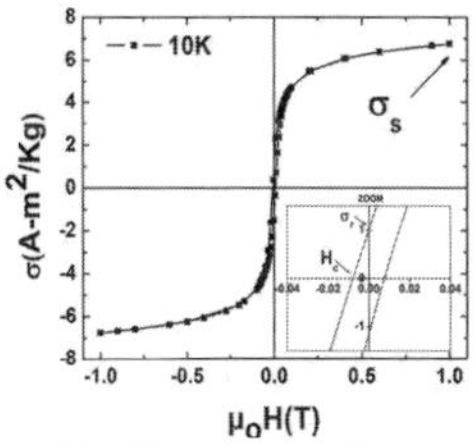

Fig. 6. Magnetization curve for nanocomposite at 10K .

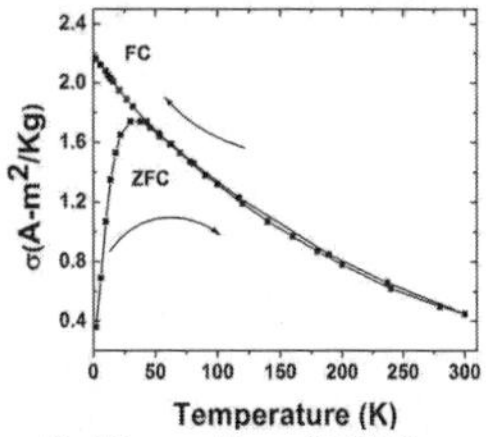

Fig. 7. Zero field cooling (ZFC) and field cooling (FC) magnetization curves at 10mT.

Comparing figure 5 and 6, it is observed a change in the magnetic behavior as function of the analysis temperature. It is important to point out, that the superparamagnetic behavior is manifested as a function of the temperature ranging from the blocking temperature (T_B) to room temperature [20]. Therefore, in order to estimate the T_B for the Iron oxide-chitosan polymer film, field cooling and zero field cooling analysis were performed at an applied magnetic field of 10mT (Figure 7). When the sample is cooled at zero magnetic fields, the total magnetization of the nanoparticles will be zero, since the magnetic moments of individual nanoparticles are randomly oriented. When temperature increases, the nanoparticles orient their magnetic moment with the external field and the total magnetization increases reaching a maximum at 30K, which is the value of the blocking temperature for the nanocomposite. At this temperature, the thermal energy become comparable to the energy gained by aligning the nanoparticle magnetic vector in the weak field and the transition from ferromagnetic to superparamagnetic behavior is observed [21]. When all nanoparticles are at the superparamagnetic relaxation state, above T_B, their magnetization follows the Curie law decreasing with increasing temperature [22]. In the case of field cooling, magnetization monotonically increases as the temperature decreases because the nanoparticles are cooled from room temperature under a magnetic field and the magnetization direction of all the nanoparticles is frozen in the field direction. The magnetization shows a maximum at 2K in the field cooling process, see figure 7. The splitting between ZFC and FC curves just above T_B has been interpreted by Osaka et al. [23] as a tight particle size distribution; this behavior is in agreement with the distribution of particle size of iron oxide nanoparticles showed in TEM micrograph.

Conclusions

The synthesis of maghemite nanoparticles from homogenous Fe(II)-chitosan polymer film was demonstrated. The nanoparticles with nearly spherical morphology and with an average size of 5 nm were embedded in the polymer matrix. The analysis of the magnetic properties of the Iron oxide-chitosan polymer film shows a superparamagnetic behavior at room temperature. Our results suggest that in situ precipitation of nanoparticles in the salt-polymer film is a promising route to the production of stabilized iron oxide nanoparticles.

Acknowledgements

The authors wish to thank R. Escudero and Marco Garza for their technical assistance in the magnetic measurements and their valuable comments. This project was financially supported a by the Consejo Nacional de Ciencia y Tecnología (CONACYT) through project No 84041.

References

[1] R.F. Ziolo, E.P. Giannelis, B.A. Weinstein, M.P. O'Horo, B.N.Ganguly, V. Mehrotra, M.W. Russell and D.R. Huffman: Science Vol. 257 (1992), p. 219

[2] K. Woo, J. Hong, S. Choi, H.W. Lee, J.P. Ahn, C.S. Kim and S.W. Lee: Chem. Mater Vol. 16 (2004), p. 2814

[3] M. Hasegawa, H. Yanagihara, Y. Toyoda, E. Kita and L. Ranno: J. Magn. Magn. Mater Vol. 310 (2007), p. 2283

[4] A. Millan, F. Palacio, A. Falqui, E. Snoeck, V. Serin, A. Bhattacharjee, V. Ksenofontov, P. Gütlich and I. Gilbert: Acta. Mater Vol. 55 (2007), p. 2201

[5] R.Y. Hong, H.P. Fu, D.Q. Di, Y. Zheng and D.G. Wei: Mater. Chem. Phys Vol. 108 (2008), p. 132

[6] S. Mornet, S. Vasseur, F. Grasset, and E. Duguet: J. Mater. Chem Vol. 14 (2004), p.2161

[7] M. Arruebo, R. Fernández-Pacheco, M. Ricardo Ibarra, and J. Santamaría: Nanotoday Vol. 2 (2007), p.22

[8] K.M. Spiers, J.D. Cashion, and K.A. Gross: Key Eng. Mater Vol. 254-256 (2004), p.213

[9] G. Ennas, A. Musinu, G. Pic(ICCD 2000)aluga, D. Zedda, D. Gatteschi, C. Sangregorio, J.L. Stanger, G. Concas and G. Spano: Chem. Mater Vol. 10 (1998), p.495

[10] J.R. Jeong, S.C. Shin, S.J. Lee and J.D. Kim: J. Magn. Magn. Mater Vol. 286 (2005), p.5

[11] J.A. López Pérez, J. Mira, J. Rivas and S.W. Charles: J. Phys. Chem. B Vol. 101 (1997), p.8045

[12] D.F.Gorozhankin, A.A. Eliseev, K.S. Napol`skii, A.V. Lukashin, A.V. Knot`ko, Y.V. Maksimov, I.P. Suzdalev, P. Görnert and Y.D. Tret'yakov: Doklady Chemistry Vol. 396 (2004), p.132

[13] S. Hashimoto, T. Uwada, H. Masuhara and T. Asahi: J. Phys. Chem. C Vol. 112 (2008), p. 15089.

[14] S. Sepúlveda-Guzmán, L. Lara, O. Pérez-Camacho, O. Rodríguez-Fernández, A. Olivas and R. Escudero: Polymer Vol. 48 (2007), p.720
[15] A.L.P. Fernandes, W.A. Morais, A.I.B Santos, A.M.L Araújo,D.S. dos Santos, F.J. Pavinatto, O.N. Oliveira Jr., T.N.C. Dantas, M.R. Pereira and J.L.C. Fonseca: Colloid. Polym. Sci Vol. 284 (2005), p.1
[16] S. Rajendran, M. Sivakumar and R. Subadevi: Solid State Ionic Vol. 167 (2004), p. 335
[17] R. Matsuno, K. Yamamoto, H. Otsuka, A. Takahara: Macromolecules Vol. 37 (2004), p.2203
[18] M.A. Garza, M.H. Hinojosa and V.A. González González: Ingenierías Vol. 9, (2006) p.14
[19] S. Gyergyek, M. Huskic, D. Makovec and M. Drofenik: Colloids and Surfaces A: Physicochem. Eng. Aspects Vol. 317 (2008), p.49
[20] M.D. Mukadam, S.M. Yusuf , P. Sharma and S.K. Kulshreshtha: J. Magn. Magn. Mater Vol. 272-276 (2004), p.1401
[21] S. Si, C. Li, X. Wang, D. Yu, Q. Peng and Y. Li: Crystal Growth & Design Vol. 5 (2005), p.391
[22] A.J. Rondinone, A.C.S Samia and Z.J. Zhang: J. Phys. Chem. B Vol. 103 (1999), p.6876
[23] T. Osaka, T. Matsunaga, T. Nakanishi, A. Arakaki, D. Niwa and H. Iida: Anal. Bioanal. Chem Vol. 384 (2006), p.593

Materials Science Forum Vol. 644 (2010) pp 57-60
© (2010) Trans Tech Publications, Switzerland
doi:10.4028/www.scientific.net/MSF.644.57

Synthesis and characterization of branched gold nanoparticles

Alvaro Mayoral[1,a], Alma Vazquez-Duran[1,b], Stefan G. Heinze[1], Miguel Jose-Yacaman[1,c]

[1]Department of Physics and Astronomy, The University of Texas at San Antonio One UTSA Circle, San Antonio, TX78249 USA

[a]alvaro.mayoralgarcia@utsa.edu, [b]alma.vazquezduran@utsa.edu, [d]miguel.yacaman@utsa.edu

Keywords: Nanoparticles, electron microscopy.

Abstract. In this work we present the synthesis of branched gold nanoparticles based on the seed mediated growth technique. The materials have been characterized by advanced electron microscopy techniques in order to elucidate morphology, size and internal structure.

Introduction

Noble metal nanoparticles such as gold, silver, platinum or palladium have been the focus of great interest due to their unique optical, catalytical, electronic and biomedical properties[1-3]. Small changes in shape, size or internal structure have a huge impact on their final properties, therefore a great effort is being made in order to control the morphology and develop new synthesis routes. Gold nanoparticles with controllable shapes are in particular of a great interest due to their unique optical properties known as localized surface Plasmon resonance (LSPR)[4] which make Au a suitable material for different applications such as enhance surface Raman scattering (SERS)[5, 6], biological and chemical sensing[7] or cancer diagnosis and treatment[2, 8-12]. The LSPR can be tuned by controlling the shape of the synthesized materials, therefore particles with peaks or branches are highly desired. Long peaks can work as antennas where electromagnetic effects are amplified. However, morphologies such as "stars" or "branched" nanoparticles[13-20] have been successfully synthesized only recently.
One of the most widely methods employed to obtain these materials has been the seed mediated growth technique[21, 22], in which a cationic surfactant (commonly Cetyl trimethylammonium bromide, CTAB) is added as a capping agent in order to control the morphology of the nanoparticles grwon. Every aspect of the synthesis such as temperature, time, additives or capping agents employed must be very carefully controlled in order to obtain new and reproducible materials. In the following, we report on our synthesis of gold nanoparticles by a seed mediated growth technique combining two surfactants, a block copolymer F127 [(poly(ethylene oxide)$_{106}$ (poly(propylene oxide))$_{70}$ (poly(ethylene oxide)$_{106}$] and CTAB, producing a high yield of branched nanoparticles.

Experimental

Preparation of Silver Seeds. 25 mL aqueous solution of 2.5×10^{-4} M $AgNO_3$ was mixed with 25 mL of 2.5×10^{-4} M Trisodium Citrate. Subsequently, 0.6 mL of 0.1M $NaBH_4$ was rapidly added to this solution (an immediate color change to a dark-green/yellow was observed). The silver solution was kept under static conditions for few hours before being used for nanoparticle growth.
Preparation of Gold Nanoparticles. An aqueous solution of 1.25×10^{-3}M CTAB (10 mL) was heated and stirred until CTAB was completely dissolved. Immediately after removing the heat source, 5 mL of 2.5×10^{-4}M $HAuCl_4$ was added, forming an orange-red mixture. To this mixture 10 mL of 1.25×10^{-4} M F-127 Pluronic were added and let under vigorous stirring for 5 minutes. The gold ions were reduced with 0.7 mL of ascorbic acid, 1M, followed by the addition of a 12.5 μL solution of silver seeds. The finished nanoparticles were then stirred for at least 30 minutes.
Characterization. Transmission electron microscopy (TEM), was performed using a Philips Tecnai G^2, 200Kv, TEM equipped with Schottky-type field emission gun, ultra-high resolution pole piece

(Cs=0.5 mm), and a scanning transmission electron microscope (STEM) unit with high angle annular dark field (HAADF) detector operating at 200 kV. Scanning electron microscopy (SEM) was performed in a FEG Hitachi S-5500 ultra high resolution electron microscope (0.4 nm at 30kV) with BF/DF Duo-STEM detector.

Results and discussion

Among the different applications of gold nanoparticles, cancer diagnosis and treatment has attracted the most attention. For this purpose branched in crystals should be created as they shift their radiation absorbance close to the infrared zone. At this wavelength human tissue and blood are transparent to the radiation but gold nanoparticles become bright making them very suitable materials which can be employed as markers. Seed mediated growth technique has been previously employed to synthesize star-shaped gold nanoparticles in the presence of CTAB[13], which absorbed radiation at ~ 800nm, however, increasing that value would be desirable and such effect can be achieved by elongating the peaks of the crystals. Figure 1a shows a SEM overview of the as-synthesized product, where most of the material presents homogeneous morphology with long branches of about 100 nm and full sizes that can reach up to 300 nm (Fig. 1b).

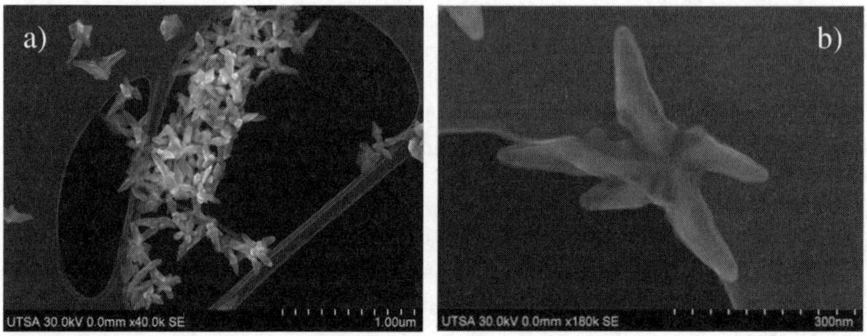

Fig. 1 SEM images of branched Au nanoparticles.

The internal structure of the particles was studied by high resolution transmission electron microscopy (HRTEM). Fig. 2 shows the TEM analysis carried out on a gold "nanostar" which developed only one branch of ~100 nm. Fig. 2b presents the weak beam dark field (WBDF) transmission electron image using the (220) spot, where contour lines, representing the thickness of the particle can be observed. This technique has also been employed to observe defects in the structure as in this mode defects are commonly highlighted[13, 23, 24]. From figure2b it can be deduced that this particle presents a single crystal structure free of defects even along the branch, which was also confirmed by HRTEM analysis were the single crystallinity structure of the nanoparticle (Fig. 2c), which grew along the [111] direction, can be observed.

EDX analysis performed on the crystals (figure 2d) showed no significant difference between the results obtained on the center and on the branch, due the much higher concentration of gold respect to the silver seed.

In the work reported here it is shown how the combination of cationic and block copolymer surfactants can be used to tune the morphologies of the crystals. F-127 and CTAB, lead to the formation of branched nanoparticles, where most of the product presented similar morphologies with branches of around 100nm. The results previously obtained in similar syntheses in the absence of F-127[13] produced polyhedral nanoparticles suggesting that this surfactant play a crucial role on the formation of such morphology stabilizating[25] the nanoparticles formed and blocking certain faces facilitating a faster growth along the <111> direction in order to finally form long peaks. Clearly, F-127 has a crucial influence on the formation of gold nanoparticles. Figure 2 would represent an intermediate step into the formation of complete branched nanoparticles. In this case the original "nanostar" symmetry would be broken by the formation of a long branch.

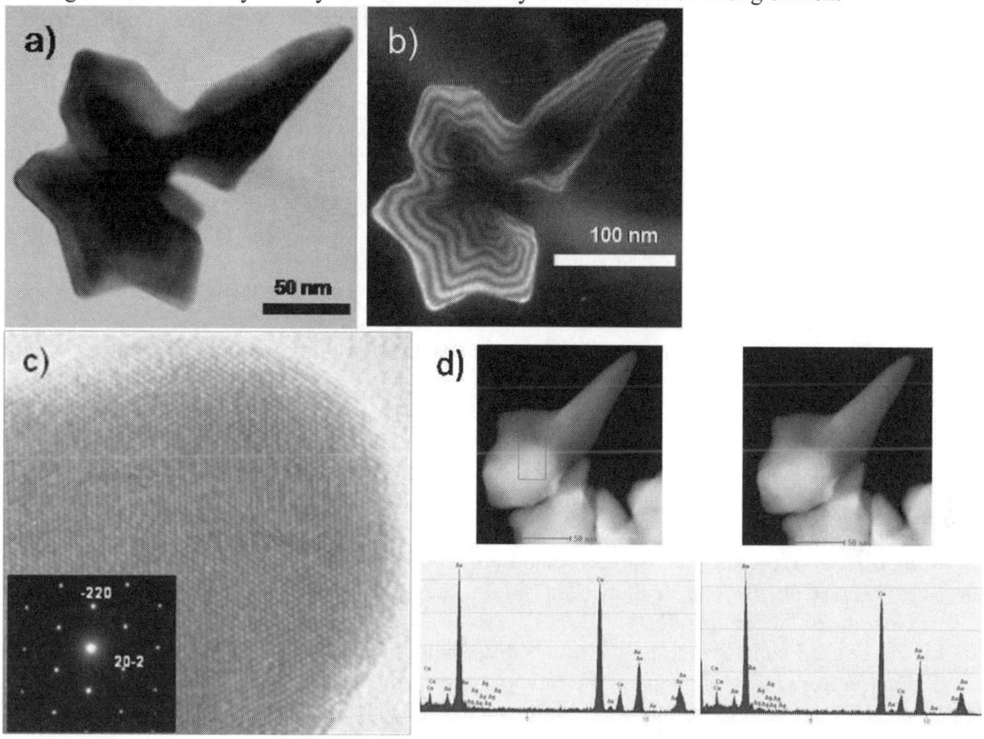

Fig. 2 a) TEM image of a branched gold nanoparticle. b) WBDF image formed by selecting the (220) diffraction spot. c) HRTEM image and selected area electron diffraction pattern (SAED) of the peak, the ED pattern is shown inset. d) EDX analyses performed on the areas squared marked where Cu peaks correspond to the microgrid used.

Summary
Gold nanoparticles have been synthesized employing the seed mediated growth technique with a pluronic surfactant added to the reaction mixture. The presence of F-127 has had a dramatic influence on the morphology of the crystals obtained, facilitating the formation of branches of about 150 nm. The material obtained presented good homogeinity proving that the combintation of both surfactant can be an alternative route for the production of branched nanoparticles, which can be potentially used as cancer markers. The transmission electron microscopy analysis revealed that some of the particles appeared to be monocrystals free of defects.

Acknowledgements
The authors would like to thank Dr. Domingo Ferrer and Dave Olmos for technical support and to the Welch Foundation, Agency project #AX-1615, NSF grant "alloys at the nano scale from the metals division Agency grant # DMR-0830074 and CONACYT (82212) for funding

References

1. Burda, C., et al., Chem. Rev., 2005. **105**: p. 1025.
2. Murphy, C.J., et al., J. Phys. Chem. B., 2005. **109**: p. 13587.

3. Rotello, V.M., Nanoparticles: building blocks for nanotechnology. 2004, New York: Kluwer Academic/Plenum Publishers.
4. Hu, M., et al., Gold Nanostructures: Engineering Their Plasmonic Properties for Biomedical Applications. Chem. Soc. Rev., 2006. **35**: p. 1084-1094.
5. Hao, E., et al., Nano Lett., 2004. **4**: p. 327.
6. Bakr, O.M., B.H. Wunsch, and F. Stellacci, Chem. Mater., 2006. **18**: p. 3297.
7. Haes, A.J., et al., Nano Lett., 2004. **4(6)**: p. 1029.
8. Eustis, S. and M.A. El-Sayed, Chem. Soc. Rev., 2006. **35**: p. 209.
9. El-Sayed, M.A., Acc. Chem. Rev., 2004. **37**: p. 326.
10. El-Sayed, M.A., Acc. Chem. Rev., 2005. **34**: p. 257.
11. Sun, L., D. Liu, and Z. Wang, Langmuir, 2008. **24**: p. 10293.
12. Perrault, S.D., et al., Mediating Tumor Targeting Efficiency of Nanoparticles Through Design. Nano Lett., 2009. **9**: p. 1909.
13. Mayoral, A., et al., Polyhedral shaped gold nanoparticles with outstanding near infrared light absorption Appl. Phys. A, 2009. **97**: p. 11-18.
14. Kuo, C.H. and M.H. Huang, Langmuir, 2005. **21**: p. 2012.
15. Wu, H.Y., M. Liu, and M.H. Huang, J. Phys. Chem. B, 2006. **110**: p. 19291.
16. Khoury, C.G. and T. Vo-Dinh, J. Phys. Chem. C, 2008. **112**: p. 18849.
17. Chen, S., et al., Monopod, Bipod, Tripod, and Tetrapod Gold Nanocrystals. J. Am. Chem. Soc., 2003. **125**: p. 16186.
18. Nehl, C.L., H. Liao, and J.H. Hafner, Nano Lett., 2006. **6**: p. 683.
19. Burt, J.L., et al., Beyond Archimedean solids: Star polyhedral gold nanocrystals. J. Cryst. Growth, 2005. **285**: p. 681.
20. Li, Z., et al., Facile Synthesis of BranchedAu Nanostructures by Templating Against a Self-Destructive Lattice of Magnetic Fe Nanoparticles. Angew. Chem. Int. Ed., 2008. **47**.
21. Jana, N.R., L. Gearheart, and C.J. Murphy, Wet Chemical Synthesis of High Aspect Ratio Cylindrical Gold Nanorods. J. Phys. Chem. B, 2001. **105**: p. 4065.
22. Jana, N.R., L. Gearheart, and C.J. Murphy, Chem. Commun., 2001: p. 617.
23. Heinemann, K., et al., J. Cryst. Growth, 1979. **47**: p. 177.
24. Elechiguerra, J.L., J. Reyes-Gasga, and M. Jose-Yacaman, The role of twinning in shape evolution of anisotropic noble metal nanostructures. J. Mater. Chem., 2006. **16**: p. 3906-3919.
25. Rahme, K., et al., Pluronics-Stabilized Gold Nanoparticles: Investigation of the Structure of the Polymer–Particle Hybrid. ChemPhysChem, 2008. **9**: p. 2230-2236.

Materials Science Forum Vol. 644 (2010) pp 61-64
© (2010) Trans Tech Publications, Switzerland
doi:10.4028/www.scientific.net/MSF.644.61

Surface modification of ZnO nanoparticles

Betancourt Galindo R.[a], Berlanga Duarte M. L.[b], Puente Urbina B.[c], Rodríguez Fernández O[d] and Sánchez-Valdes S.[e]

[1]Centro de Investigación en Química Aplicada, Blvd. Enrique Reyna #140, Saltillo, México

[a]rgalindo@ciqa.mx, [b]lydiab@ciqa.mx

Keywords: ZnO nanoparticles, surface modification, chemical hydrolysis

Abstract. The synthesis and surface modification of ZnO nanoparticles, achieved through a controlled precipitation reaction using NaOH as precipitating agent with a 0.2 molar ratio were studied. The surface modification was carried out through a chemical hydrolysis reaction at 80°C for 6 h, in order to promote better adhesion between inorganic particles and a polymer matrix when they were used in composite materials. Both 3-aminopropyltrimethoxysilane (APTMS) and difunctional alcohol spiroorthocarbonate (SOC DIOL) were used as modifying agents in a 0.5:1 ratio with respect to ZnO nanoparticles. According to x-ray diffraction (XRD), electron transmission microscopy with scanning modulus (STEM) and infrared spectroscopy (FTIR), the ZnO crystalline structure did not change after treatment. Surface modification was performed in order to subsequently use ZnO nanoparticles in medical antimicrobial devices.

Introduction

The inorganic nanoparticles of ZnO have attracted wide attention due to the antibacterial properties they possess. A large number of methods have been developed to synthesize ZnO nanoparticles with controlled size and morphology. These methods are: hydrothermal proccss [1], sol-gel [2] and precipitation [3], among others. Recently, the synthesis of inorganic ZnO nanoparticles has found potential application in many fields such as optical sensors, electronics [4], and biotechnology [5]. However, these nanoparticles tend to agglomerate due to their great superficial area and high superficial energy. Thus they show poor dispersion and need to be modified superficially. Different methods exist for the superficial modification of the nanoparticles, such as plasma modification, chemical hydrolysis, and treatment with agents like titanates, oleates, siloxanes, among others [4]. Several authors [6] have reported the surface modification of ZnO nanoparticles using aminopropyltriethoxysilane (APTES) under various environments. They reported that it was possible to control the coating which acted as a growth inhibitor even at higher temperatures. Other authors [7] have reported the ZnO ultrafine particles synthesis by in situ modification method, using zinc carbonate hydroxide (ZCH) as the precursor which was modified in situ with silica and trimethyl silozane (TMS). They reported that the photocatalytic activity of the ZnO particles was greatly suppressed after surface modification and that these modified particles had good UV shielding properties compared with the non-modified ZnO particles.

This work reports the superficial modification of ZnO nanoparticles by chemical hydrolysis using the APTMS and SOC DIOL as modifying agents in a ratio of 0.5:1 with respect to ZnO nanoparticles. The results of FTIR and the STEM showed that when the nanoparticles were superficially modified they did not suffer any change in the crystalline structure.

Experimental Part

Materials. Zinc acetate ($Zn(O_2CCH_3)_2$), Sodium Hydroxide (NaOH) and octane 98 % from Aldrich were used for the synthesis of ZnO nanoparticles. 3-aminopropyltrimethoxysilane (APTMS) 97 %

from Aldrich was used for the superficial modification of the nanoparticles. The difunctional alcohol spiroorthocarbonate (SOC DIOL) was synthesized in our lab (CIQA) [8].

Synthesis of ZnO nanoparticles. The preparation of ZnO nanoparticles was carried out by precipitation using 0.1 M $Zn(CH_3COO)_2$ and 0.1 M NaOH. The reaction was prepared from solution of zinc acetate in ethanol and they were refluxed for 3 h at 80°C. The resulting product was washed, centrifuged and vacuum dried at 80°C for 4 h.

Superficial Modification of ZnO nanoparticles. The superficial modification of ZnO nanoparticulas was carried out using a molar ratio of 0.5:1 of modified agents (APTMS and SOC DIOL) to nanoparticles. In both cases, octane was used as the dispersion agent. This reaction was carried out at 80 ºC with constant agitation and refluxed for 6 h. Once the system was cooled, the modified material was decanted and dried for 12 h at 80 ºC for its subsequent characterization.

Characterization. The analysis with XRD was carried out in a diffractometer SIEMENS D-5000 with a CuKα radiation to identify the crystalline phase present in the obtained ZnO nanoparticles. Morphology and superficial modification of the nanoparticle samples were analyzed with a scanning electron microscopy (SEM) microscopy with a STEM modulus. The STEM observations were performed with a Jeol-JSM7104F with a STEM modulus and a field emission gun at an accelerating voltage of 200 kV in a JEOL JFM7401 F equipment. The FTIR analysis was carried out with a Nicolet Magna 5500 spectrometer using the preparation of KBr tablets with the obtained sample to identify the absorption bands for the Zn-O bond and the corresponding bands of the modification agents.

Results and Discussion

Figure 1 shows the XRD patterns of ZnO nanoparticles obtained from precipitation. The peaks of these XRD patterns are in good agreement with wurtzite crystal structure and they also correspond with the JCPDS 36-1451. This means that the molar concentrations used for the synthesis of the nanoparticles are adequate.

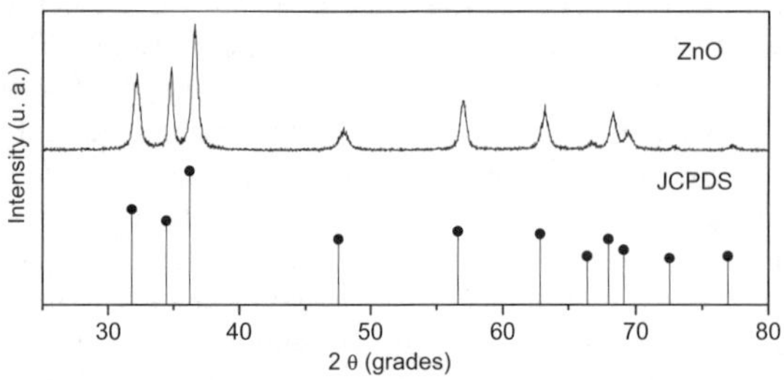

Figure 1. XRD patterns of ZnO nanoparticles.

The average particle size of ZnO nanoparticles was determined using the following Scherrer equation. Since the peak width was dependent on the crystal size and on the heterogeneous deformation in the network:

$$d = \frac{0.9\,\lambda}{\beta \cos \theta}$$

Where λ = 1,5406Å; β is the average peak width and θ is the diffraction angle.

The average ZnO particle size was calculated to be about 23.44 nm confirming that it was possible to obtain ZnO nanoparticles using this precipitation technique.

The infrared spectra of the ZnO nanoparticles with and without modifier are showed in Figure 2. In the spectrum of the un-modified ZnO nanoparticles the corresponding band related to the stretching of the Zn-O bond appears at 439 cm^{-1}. The spectrum of the modified ZnO nanoparticles with both modifier agents corroborates that these nanoparticles are superficially covered by a modifier, since the typical absorption bands of both superficial modifier agents are present. In addition it is observed that a change in the crystal structure of the nanoparticles does not exist, as it will be confirmed later by the STEM analysis.

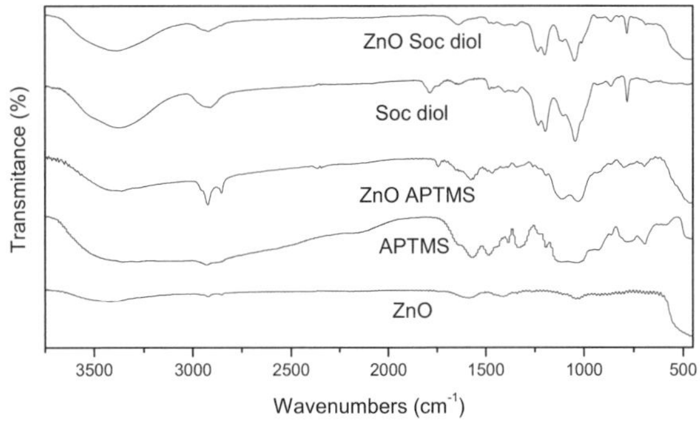

Figure 2. FTIR Spectra of ZnO nanoparticles without modifying and modified

Figure 3 shows the STEM micrographs of the modified ZnO nanoparticles with: a) APTMS and b) SOC DIOL. Both micrographs clearly show the presence of agglomerates and a more or less homogeneous layer covering the nanoparticles. The result means that even though the nanoparticles are agglomerated, they are superficially modified with no adverse effect on their dispersion when they are incorporated in any polymer matrix. Because of the noticeable agglomerates, it was not possible to determine the average particle size from these images.

a) b)

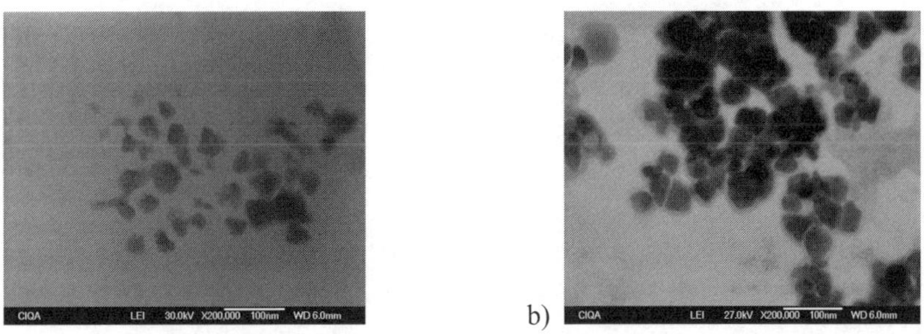

Figure 3. STEM micrographs of ZnO nanoparticles after being modified with the two agents: a) APTMS and b) SOC DIOL.

Conclusions

The synthesis of ZnO nanoparticles was successfully achieved in our lab by a controlled precipitation reaction. Furthermore, the surfaces of these nanoparticles were also successfully modified using APTMS and SOC DIOL. Agglomerates were present on the coated nanoparticles, however these had no adverse effect on the crystalline structure or their dispersion property. Our superficially surface-modified ZnO nanoparticle would prove to be of potential application in composite materials in the fields of optical sensors, electronics and biotechnology.

Acknowledgments

The authors are grateful to CONACYT-México for the financial support of projects: CB-2007/80159 and CB-2007/80108.

References

[1] Pookmanee P., Kidrob W., Intarapoom R., Phanichphant S. *Journal of Microscopy Society of Thailand.* 20 (1) 72-78 (2006).
[2] Meulenkamp E. A. *J. Phys. Chem. B*, 102, 5566-5572, (1988).
[3] Chen C. Liu P. Lu C. *Chem. Engineering Journal* 144 509-513 (2008).
[4] Honh R. Y., Li J. H., Chen L. L., Lui D. Q., Li H. Z., Zheng Y., Ding J. *Power Technology* (2008).
[5] Guo L., Yang S., Yang Ch., Yu P., Wang J., Ge W. and Wong G. *Chem Mater.* 12, 2268-2274 (2000).
[6] Grasset F., Saito N., Li D., Park D., Sakaguchi I., Ohashi N., Haneda H., Roisenel T., Mornet S. and Duguet E. *J of Alloys and Compounds.* 360, 298-311 (2003).
[7] Cao Z., Zhang Z., Wang F., and Wang G., *Colloids and Surfaces A: Physicochem. Eng. Aspects.* 340, 161-167 (2009).
[8] Ricardo Acosta Ortiz *, Maria Lydia Berlanga Duarte, Marco Sangermano, Amy Grace Savage Gomez, Aida E Garcia Valdez "*Novel diol spiro orthocarbonates derived from glycerol as antishrinking additives for the cationic photopolymerization of epoxy monomers*": Polymer International (in press 2009)

Materials Science Forum Vol. 644 (2010) pp 65-68

doi:10.4028/www.scientific.net/MSF.644.65

Topographical characterization of electrodeposited nickel nanoparticles on an Indium Tin Oxide on glass thin film

Gerardo T. Martinez[1, a], Genaro Zavala[2,b] and Marcelo Videa[3,c]

[1]Laboratorio de Microscopia. Centro de Investigación en Química Aplicada. Blvd. Enrique Reyna No. 140, Saltillo, Coah. 25253, México

[2]Departamento de Física. Tecnológico de Monterrey, Campus Monterrey, Av. Garza Sada Sur 2501, Monterrey, N.L. 64849, México

[3] Departamento de Química. Tecnológico de Monterrey, Campus Monterrey, Av. Garza Sada Sur 2501, Monterrey, N.L. 64849, México

[a]gtadeo@ciqa.mx, [b]genaro.zavala@itesm.mx, [c]mvidea@itesm.mx

Keywords: Electrodeposition, nickel nanoparticles, characterization, atomic force microscopy.

Abstract. Topographical characterization of electrodeposited nickel nanoparticles was carried out by Atomic Force Microscopy (AFM) and Scanning Electron Microscopy (SEM). The particles showed a hemispherical growth, typical from a Volmer-Weber growth mechanism. However, they presented different morphology and crystallization types under similar electrochemical conditions. Preferential growth on determined sites on the substrate surface at different electrochemical conditions was also studied, showing that the particle growth strongly depends on the competition of different electrochemical processes, such as Hydrogen Evolution Reaction (HER), which are promoted preferably by the pulse intensity.

Introduction

The development on magnetic storage media is improving very fast. The recording densities have increased from 60% to 100% every year, for the past 20 years, reaching recording densities in the order of Gbit/cm²[1]. This development demonstrates the importance of finding better means to storage information, because in some cases, the limitations of these technologies start to show difficulties. Nowadays, new technologies and suitable materials that allow the continued increase in the information density per unit area are needed. This implies the reduction of the bit size, and in consequence, the use of materials with suitable magnetic properties.

New manufacturing methods for industrial applications, which are inexpensive and easy to implement for mass production, are in demand. In the field of electrochemistry, there are processes that allow the deposition of nanoparticles and the growth of magnetic nanowires at low cost, high control of parameters, and certainly fast. It has been demonstrated that, with carefully designed electrochemical experiments, cobalt magnetic nanowires can be fabricated [2]. In the same way, conditions of current and voltage have been established for the electrodeposition of Ni nanoparticles according to a design of desired size and distribution [3]. The electrochemical methods of synthesis offer a suitable alternative for the fabrication of magnetic elements with magnetic properties for long-term storage information and may fulfill the future needs.

Electrodeposition of metallic particles through a double potential pulse technique has been studied extensively [4-5]. The mechanism of nucleation and growth has also been studied and it is concluded that depending on the overpotential pulse, the nucleation can be either instantaneous or progressive [6-9].

In this work we explore the use of a single current pulse, which forces the potential of the system to adjust to the imposed kinetics for electron transfer. The behavior of the system potential is similar to the one imposed by the double potential pulse technique. The results obtained on the synthesis of nickel particles through a current pulse method are reported. Topographical characterization of the electrodeposited particles is presented, as a basis for further magnetic characterization and consideration as a suitable material for magnetic data storage.

Experimental Setup

The electrodeposition experiments were carried out in a three electrode cell, with a thin film of Indium Tin Oxide (ITO) on glass as working electrode (WE), a Pt electrode as counter electrode (CE), and an Ag/AgCl (RE) as reference electrode. A 0.05M $NiSO_4$ with 0.1M Na_2SO_4 solution was used. The pulses were applied using a 273A EG&G Princeton Applied Research potentiostat/galvanostat. Several types of experiments were performed:

- Equal intensity pulsed current experiments of 0.5 mA with different durations: 90 ms, 110 ms, 500 ms, 1000 ms, 5000 ms, 10000 ms and 15000 ms.
- Equal period of time experiments of 55 ms with different intensity pulsed current: 0.5 mA, 1.0 mA, 1.5 mA, 3.0 mA, 4.0 mA, 5.0 mA and 6.0 mA.
- Equal transferred charge experiments of 40 μC, varying the intensity pulsed current and period of time.

The samples were topographically characterized using a Veeco Nanoscope IIIa Multimode Atomic Force Microscope with Veeco MESP silicon tips in tapping mode and a LEO Field Emission Scanning Electron Microscope.

Results and Discussion

The electrochemical results are discussed elsewhere [10]. For the morphology of the deposited particles, it was found to have different crystallization types, which are:

- peaked granular shaped
- smooth granular shaped
- big flakes shaped
- small flakes shaped

Fig. 1 shows these different types of crystallization by SEM micrographs. The morphology of the nanoparticles was found to be independent of the variables such as period of time, intensity of pulsed current or transferred charge, since the different types of morphology were found in both low intensity and high intensity pulsed current experiments or for both short and long time experiments. The different types of morphology of the particles were found in the same experiment, which supports the idea that particle morphology is affected by other variables from those discussed above, since all these particles are grown in the same electrochemical conditions. It can be observed that the particle size is not a parameter to consider either for the different crystallization types, since in Fig. 1 particles of similar sizes but with different crystallization types are formed, such as smooth and flaked surface particles.

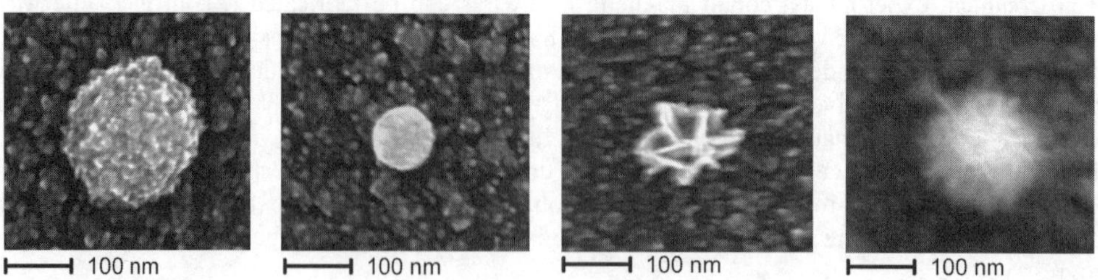

a) b) c) d)

Fig. 1: The morphology of the particles was found to have different crystallization types: (a) peaked granular shaped, (b) smooth granular shaped, (c) big flakes shaped and (d) small flakes shaped. Images show SE signal using 10 kV acceleration voltage.

The growth of nanoparticles onto the sample was found to be related to the pulsed current intensity. For low intensities, the particles were found to deposit on irregularities of the substrate, such as scratches. These zones of the surface are considered to be low potential energy places, where the reduced atoms find suitable conditions to nucleate. Fig. 2a shows the results for a low intensity pulsed current experiment (1.5 mA - 55 ms), where a higher density of particles deposited preferably at a scratch of the surface rather than in other areas. The particles in this zone are bigger than most of the other particles, indicating the possibility of earlier nucleation and/or faster growth.
For an experiment in which a high intensity pulsed current was used, the deposition pattern was found to be the opposite to that of low intensity pulsed current, since the particles do not grow on the irregularities of the surface. In this case, the particles nucleate and grow in the regular surface of the sample. Fig. 2b shows the particle deposit of a very intense pulsed current experiment (30 mA - 1.33 ms), where it can be observed that the particle deposition avoids the irregularities of the sample. The image shows a very dense deposit composed of small particles similar to a thin film.
For low intensity current pulses, the rate of charge transfer will be in general slow except in certain active places, like scratches, that allow the reduction and nucleation of the Ni atoms. On the other hand, for the case of high intensity current pulses, the polarization of the electrode may favor the Hydrogen Evolution Reaction (HER) which is known to compete with the Ni reduction [11]. Therefore, it is reasonable to expect that HER will mostly occur on the irregularities of the sample blocking these area to Ni deposition and leaving the rest of the surface of the electrode for the deposition of Ni particles.

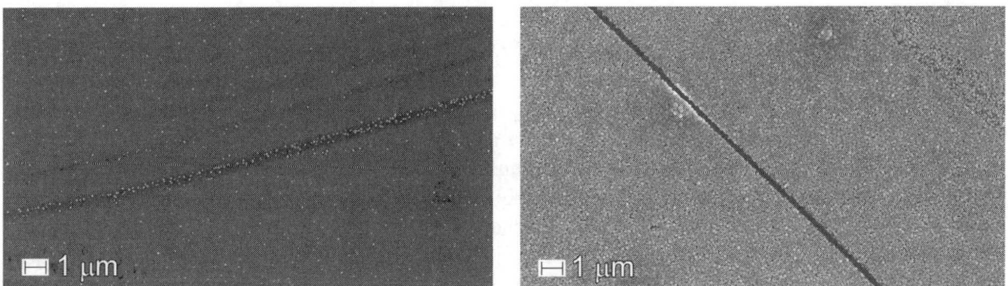

a) b)

Fig. 2: a) The particle deposition was found to be preferable on irregularities at the surface, such as scratches, when the pulsed current intensity was low. The image was taken from the experiment when a 1.5 mA current pulse was applied for 55 ms. b) The particle deposition for very intense current pulses was found to be preferable on the regular surface of the sample, avoiding the irregularities, such as scratches due to the competing HER reaction. This image was taken when a 30 mA pulsed current is applied for 1.33 ms. Both images show SE signal using 10 kV acceleration voltage.

The AFM characterization of the particles, as it provides 3-dimensional information of the sample, gave information about the aspect ratio of the particles. The aspect ratio is the relation between height and diameter of the particle. It must be 0.5 to be perfectly hemispherical. The particles found in the deposit had the characteristics shown in Figure 3. It can be observed that the particle aspect ratio remains near the 0.5 value in spite of the particle size. The particles have a hemispherical shape, but the bigger particles (300-500 nm) tend to be flatter vertically than horizontally, since its aspect ratio is smaller than 0.5. Moreover, the particles of a smaller size (30-50 nm) tend to be the opposite, flatter horizontally than vertically, since its aspect ratio is greater than 0.5.However, in general it can be assumed that the particles grow in a hemispherical shape.

AFM micrographs of low density and high density particle deposits on the sample showed that the morphology of the particles is hemispherical in spite of the particle density. The AFM analysis also showed that the growth mechanism of the particles is the Volmer-Weber (Island) growth regardless of the speed of charge transfer or particle density deposit, since the particles maintain a hemispherical shape.

Figure 3: Particle aspect ratio found by AFM characterization with AFM micrograph of the particles showing hemispherical shape.

Diameter [nm]	Aspect ratio
500	0.38
400	0.39
300	0.39
200	0.52
100	0.47
70	0.47
60	0.45
50	0.58
40	0.68
30	0.64

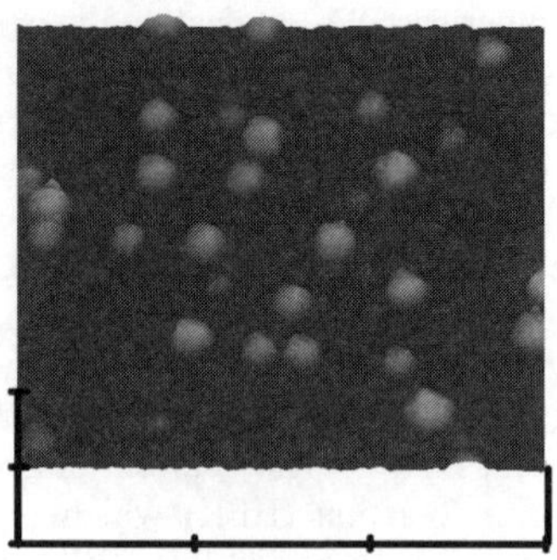

Conclusion

The morphology of the particles found by AFM and SEM inspection indicates four types of crystallization: peaked granular, smooth granular, big flakes and small flakes shape. These types of crystallization were found to be independent of pulsed current intensity, charge transfer or duration of pulsed current, since they were found to form under the same electrochemical conditions. Particle size and growth limited by diffusion are not variables to consider for the crystallization types. AFM analysis showed that the particle morphology is hemispherical in spite of the particle density, conforming that a Volmer-Weber growth mechanism is found, regardless of speed of charge transfer or particle density of the deposit. The different distribution of the deposits when a low current pulse or a high current pulse, which involves competition between the Nickel reduction and the HER, suggest a means of controlling the distribution of the nanoparticles in the deposit.

References

[1] J. Windeln, C. Bram, H.L. Eckes, D. Hammel, J. Huth, J. Marien, H. Rohl, C. Schug, M. Wahl, A. Wienss. Appl. Surf. Sci. Vol. 179 (2001) p. 167
[2] M. Darques, L. Piraux, A. Encinas, P. Bayle-Guillemaud, A. Popa, U. Ebels. Appl. Phys. Let. Vol. 86 (2005)
[3] S. Y. Chacón, in "Electrodeposición de nanopartículas para la electrocatálisis de la reacción de evolución de hidrógeno (REH)". Thesis. ITESM. (2004)
[4] E. Budevski, G. Staikov, W.J. Lorenz, Electrochim. Acta. Vol. 45 (2000) p. 2559
[5] M. Ueda, H. Dietz, A. Anders, H. Kneppe, A. Meixner, W. Plieth, Electrochim. Acta Vol. 48 (2002) p. 377
[6] B. Sharifker and G. Hills, Electrochim. Acta Vol. 28 (1983) p. 879
[7] B. Sharifker, J. Mostany, J. Electroanal. Chem. Vol. 177 (1984) 13
[8] M. Sluyters-Rehbach, J.H.O.J. Wijenberg, E. Bosco, J.H. Sluyters, J. Electroanal. Chem. Vol. 236 (1987) p. 1
[9] L. Heerman, A. Tarallo, J. Electroanal. Chem. Vol. 470 (1999) p. 70
[10] G.T. Martínez, G. Zavala, M.Videa, J. Mex. Chem. Soc. Vol 53 (2009) p.7
[11] M.P. Zach and R.M. Penner, Adv. Mat., Vol 12 (2000) p 878

Materials Science Forum Vol. 644 (2010) pp 69-74
© (2010) Trans Tech Publications, Switzerland
doi:10.4028/www.scientific.net/MSF.644.69

Analysis of nanocrystalline intermetallic compounds from their X-ray diffraction patterns

H. Rojas-Chávez[1, a], F. Reyes-Carmona[2, b], G. Plascencia[1] and D. Jaramillo-Vigueras[1]

[1]Centro de Investigación e Innovación Tecnológica-IPN, Cerrada de CECATI s/n, Col. Santa Catarina, Del. Azcapotzalco. México, D.F. 02250

[2]Facultad de Química-UNAM, Circuito de la Investigación Científica s/n, C.U., Del. Coyoacán. México, D.F. 04510

[a]hrojasc@ipn.mx, [b]fidelr@servidor.unam.mx

Keywords: High-energy ball milling, PbSe, Nanocrystalline, Solid-state chemical reactions.

Abstract. Synthesis of intermetallic PbSe induced by high-energy ball milling has been studied. Systematic analysis of transformation in the resulting phases of milling from 0 to 10 h has been traced by characterizing the microstructures in terms of morphology, crystallite size, and percentages of phases formed as a function of milling time. Results reveal the formation of two phases. Where the system of PbO-Se transforms gradually to PbSe. Complete transformation is achieved after 10 h of milling time. Study of particle structure by the Rietveld Method further corroborates the value of the nano-order crystallite diameters as evaluated from the microscopic studies.

Introduction

The scientific interest in nanocrystalline and nanostructured materials strives in the change and interesting mechanical, thermoelectric, optical properties among many others compared to bulk materials, due to the reduction of the scale length [1, 2]. The synthesis of nanostructured PbSe can be performed using different techniques. Among these, high-energy ball milling (HEBM) is a simple and efficient way to achieve this purpose [1]. In this process, the material is subjected to repeated impacts, which leads to clustering, hardening and reduction of particles sizes onto nanometric scale; it may also induce plastic deformation [1, 2].

The synthesis induced by HEBM of mixtures of crystalline powders is attributed to solid-state chemical reactions, the kinetics of which is controlled by the excess point and lattice defects [3-6]. Recently, gas-solid reactions induced by HEBM between metal powders and a gaseous atmosphere have received much attention. In such cases, reaction products most usually have a nanocrystalline structure [7, 8]. Typical examples are oxidation reactions that are easily induced and are often difficult to prevent in the milling process. In the case of HEBM, these reactions can occur at room temperature, as long as sufficient gas is available to complete chemical transformations [8].

In the present study, PbO–Se powder mixtures were ball milled under an air atmosphere at room temperature to study the reaction sequence during milling. The PbO–Se system and the air atmosphere were chosen because of the easy of formation of lead and selenium oxides. Various analyzing techniques such as X-ray diffraction and high-resolution transmission electron microscopy were employed to investigate structural and morphological development of the as-milled materials.

Experimental procedure

Synthesis

In the present study we used commercial PbO (99.9 mass%) and Se (99.9 mass% or higher) powders from Aldrich. According to manufactures, particle size of PbO was 5 μm and the Se 50 μm. Samples of PbSe were prepared by milling PbO–Se mixtures in different composition ratios and under different processing times with a high-energy type ball mill (SPEX 8000 mixer/mill) in an air atmosphere at room temperature under the following conditions: vial and ball material, nylamid and ZrO_2; inner volume of the vial, 10 ml; ball to powder mass ratio, 10:1; milling time, up to 10 h. Powders were mixed in the initial 4:1.4 mass ratios in order to form PbSe. Milling process was carried out in a nylamid vial with hemispheric faces to prevent stagnant zones. The powder mixtures were introduced in nylamid containers under atmospheric conditions. Sampling was carried out by taking powder milled mixtures from vials at several times.

Structural characterization

The X-ray powder diffraction patterns of the samples were measured at room temperature in air in a Theta-Theta Bruker D-8 Advance diffractometer that had a Bragg–Brentano geometry, Cu Kα radiation, a graphite secondary-beam monochromator, and scintillation detector. The diffraction intensity as a function of the angle 2θ was measured between 20 and 110º, with a 2θ step of 0.02º and a counting time of 10 s per point. The crystalline structures were refined via the Rietveld method using the FullProf code [9]. Crystallite morphology was modeled using spherical harmonics as base functions [10]; assuming crystallite microstrain to be isotropic; and modeling the background with a polynomial function that, in addition to the constant, linear, quadratic, and cubic terms in 2θ, also included the terms $(1/2\theta)$ and $(1/2\theta)^2$. Phase identification was performed using the PDF file version 2.0. (PDF-2 Database, JCPDS-ICDD, Newton Square, PA, 1999).

Morphological characterization

The morphology and particle size were analyzed by high-resolution transmission electron microscopy in a FEI TECNAI G^2-F30 microscope. The as-milled powder was dispersed in ethanol before being placed in a copper grid covered with Formvar.

Results and Discussion

The structure evolution of the PbO–Se system in several stages of the mechanical treatment (0-10 h) is shown in Fig. 1. The XRD pattern of the starting as-prepared sample (0 h) is characterized by several distinct and sharp diffraction peaks that can all be ascribed to PbO (ICSD card 005-0570) and Se (card no. 086-2246) phases, Fig. 1a. No additional diffraction peaks can be observed, implying that PbO and Se are the only phases present.

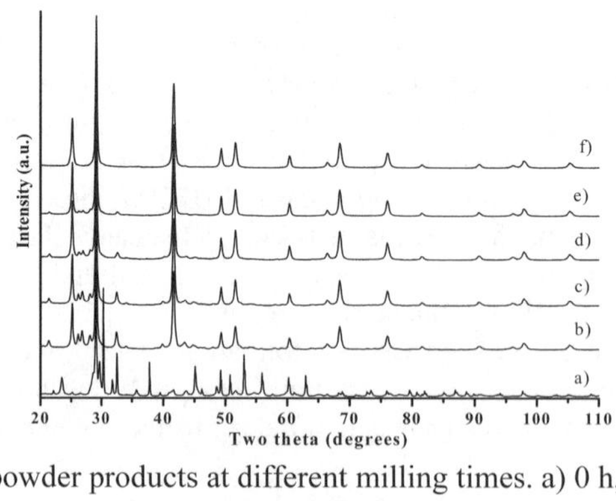

Fig. 1: XRD patterns of powder products at different milling times. a) 0 h, b) 2 h, c) 4 h, d) 6 h, e) 8 h and f) 10 h.

In the pattern, Fig. 1b – e, of the as-milled samples there are some small peaks from $PbSeO_3$ (card no. 071-0831) yet there are very intense ones which were attributed to the PbSe phase given in the ICSD card no. 006-0354. After 10 h, only PbSe phase was detected while there was no presence of another phase peaks. In Fig. 1f, shows the XRD pattern for lead selenide obtained in the desired atom ratio 1:1. As expected, no powder impurity phases were detected by XRD as shown in Fig 1.

In the PbO–Se system, the mixture of PbO and elemental Se undergoes both solid-state reactions and solid gas reactions during HEBM as clearly indicated in Fig. 1.
By taking into account the thermochemistry [11] involved through those changes expected to take place during milling, it is proposed the following scheme of reactions for the PbO–Se system:

$$Se + O_2(g) \rightarrow SeO_2;\ \Delta H_{f1} = -225.5\ kJ/mol \quad (1)$$

$$Se + O_2(g) \rightarrow SeO_2(g);\ \Delta H_{f2} = -110.5\ kJ/mol \quad (2)$$

$$PbO + SeO_2 \rightarrow PbSeO_3;\ \Delta H_{f3} = -94.5\ kJ/mol \quad (3)$$

$$PbO + Se \rightarrow PbSe + 0.5\ O_2(g);\ \Delta H_{f4} = 118.1\ kJ/mol \quad (4)$$

Thus, reactions (1) and (2) are expected to take place as a consequence of the oxygen potentials. Hence, SeO_2 either as a solid or gas reaction product can be expected. That the atmosphere oxygen potential (p_{O_2} = 0.21 atm), also promotes through reaction (3), a partial substitution of oxygen in SeO_2 to give way to the formation of $PbSeO_3$. Simultaneously to those chemical changes, a gradual reduction of oxide phases by presence of Se should occur to a point where the mayor reaction product would be the PbSe phase. While reactions products from reactions (1) and (2) were not quantitatively traced, some reports in the literature support this mechanism [11, 12]. A fact that justifies the evolution of reaction (2) is that perceived after the vial was opened to take milling products, a characteristic smelling of a gaseous phase was clearly identified.

By analyzing enthalpy values of reactions (1) – (4) one can observe that while reactions (1) – (3) are expected to occur by releasing energy gradually, the reaction (4) on the contrary has an endothermic character. Hence, as particles are continuously milled a part of that energy is stored thus inducing the formation of PbSe through reaction (4). Although previous theoretical and experimental results are congruently related, a further study is suggested to take into account entropy contributions to quantify more precisely those transformations.
It is acknowledged that high energy ball impact results in particle size reduction, down to nanometric sizes. Hence, new surface area is continuously formed during milling under ball impacts, giving rise to a very large surface area. Indeed, the newly created surfaces were readily covered by a layer of PbO in order to achieve reactions (3) and (4). Due to this fact, it was not possible to detect SeO_2 by XRD. Although reaction (4) has a positive heat of formation, it is interesting to observe that at room temperature it took place. According to thermochemical concepts ΔH_f for the reaction (4) is greater than zero. This fact reinforces the idea that the new phases formed by HEBM technique are associated with the energy involving the mechanical equivalence to that magnitude of heat produced by the milling media collisions and the energy necessary to activate migration defects.

All as-milled powders were analyzed quantitatively by Rietveld's method. In every case, the qualities of the refinements were confirmed by factors of reliability and by visual examination of the plotted difference between the measured and calculated profiles, see Fig. 2. The obtained sample is identified as a pure cubic phase (PbSe No. 006-354) with a lattice parameters a = 6.121 Å for PbSe.
The particle sizes obtained from Rietveld's refinement was found to be 20 and 25 nm. The estimated particle sizes are consistent with the average particle sizes observed from HRTEM. PbSe is cubic, space group Fm-3m with Z=4. Both Pb and Se atoms are in fixed positions at ½½½ and 000 respectively, and both atoms are in octahedral coordination.

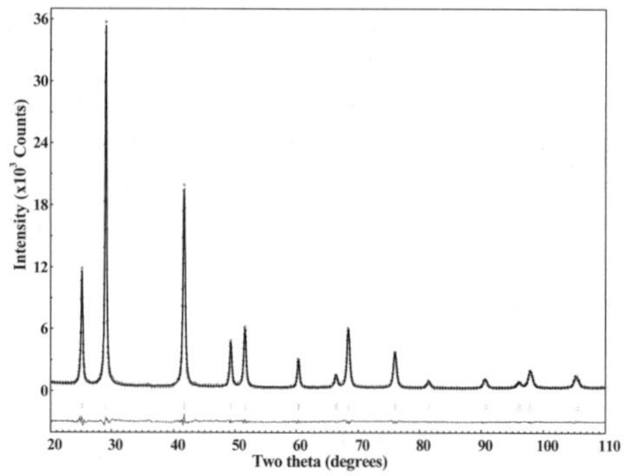

Fig. 2: Rietveld refinement plot of PbSe compound synthesized by HEBM after 10 h of milling time. In the upper curve, dots correspond to the experimental data, and the continuous line corresponds to the calculated diffraction pattern. The lower curve represents the difference between the experimental and calculated data. The upper marks correspond to lead selenide.

The particle size obtained from Rietveld's refinement can be observed in Fig. 3a, the average value of the particle size was found to be in the order of 23 nm. The sizes of the crystallites were reduced in the initial milling time from the micrometric down to the nanometric scale. After 6 h milling time the particle size increased from 23.5 to 24.3 nm. However, after that stage of milling, particle sizes decreased from 24.3 nm down to about 21.75 nm. A variation in lattice parameters for the PbSe phase is shown in Fig. 3b. The lattice parameters obtained from the Rietveld refinement were used to obtain lattice volumes of the average crystallites, see Fig. 3c. According to Swalin [13], there is a relation between the lattice volume and the heat of formation of the crystal, ΔH_f. Then a large negative value of heat of formation means that energy is evolved upon formation of the crystal. In the case of highly exothermic reactions, considerable contraction should be expected. If one take into account the reference volume of a unit cell of PbSe [14] and that is compared against those results found in the present work then as seen in Fig 3 c, a continuous transformation starts from that point. Therefore a measure of the sequence of the chemical transitions should be related to the positive enthalpy of formation of that specie.

Fig. 3d shows variation of the amount of PbSe and $PbSeO_3$ of the milled samples as a function of milling time. The PbSe content increases linearly with the milling time and attains a maximum value after 10 h milling. On the other hand, the amount of $PbSeO_3$ is decreased as milling time increases. Comparing the above results, these suggest that during the first milling stage the milling air pressure decreased in order to increase both the SeO_2 and $PbSeO_3$ powder content. At this stage the PbSe phase was the predominant one. Fig. 3d shows that the amount of the PbSe phase increases and the PbSeO3 content is drastically decreased.

By considering those results shown in Figs. 3 (a-d), one can elucidate that the rate of mechanical energy is gradually introduced into the powder system in such a way that this mechanically equivalent is transformed to an activation energy which promotes a series of chemical reactions. As expected, that energy must be dissipated congruently with that set of reactions (1-4) given previously. Hence, although the rate of chemical changes induced by the HEBM process, are identified as a rate of chemical changes, it can not be acutely disclosed in terms of energy spent to account for elasto-plastic deformation neither to account for that energy to increase the area/volume ratio of particles nor to account for diffusion of species or to quantify induced defects by atomic considerations.

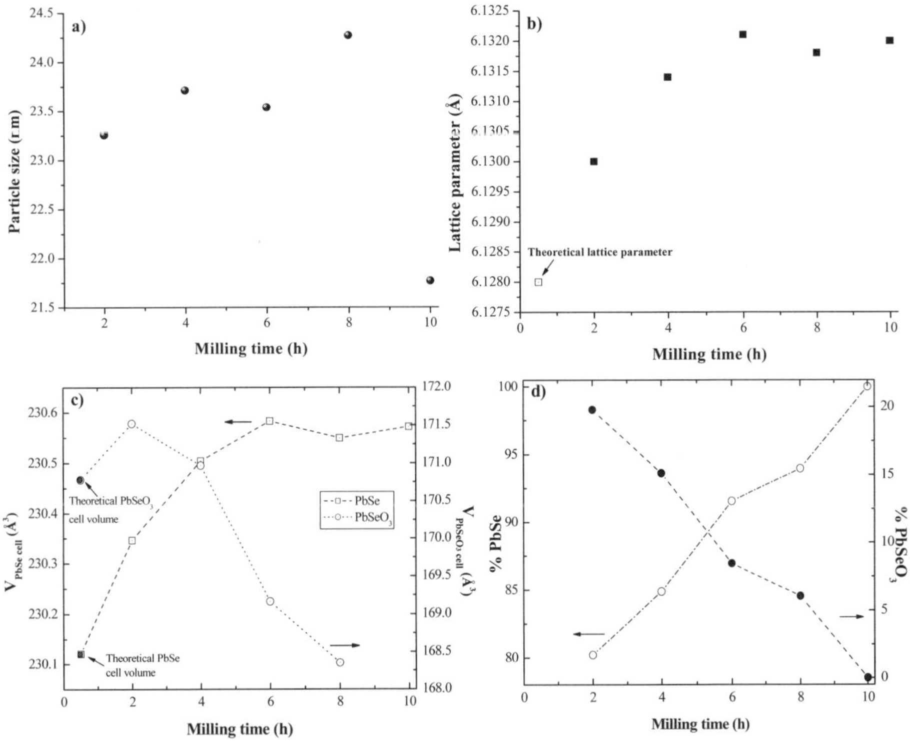

Fig. 3: The a) Particle sizes and b) Lattice parameters values of PbSe at different milling times. c) Details on the variation in the content of PbSe and $PbSeO_3$ phases at different milling times. d) Variation in the cell volume of PbSe and $PbSeO_3$ phases at different milling times.

Typical HRTEM micrographs for the powders in as-milled conditions are showed in Fig. 5. The majority of the PbSe crystallites have dimensions ranging from about 15 nm to 20 nm. After 2 h of milling the average particle size is reduced to less than 30 nm, see Fig. 5a.

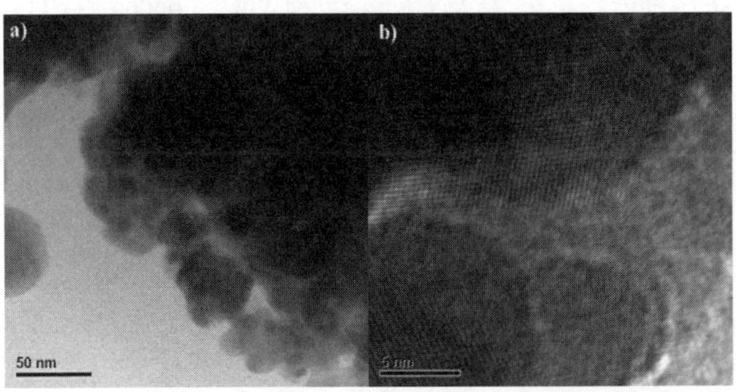

Fig. 5: HRTEM images of PbSe powders produced by high-energy ball milling after 10 h.

According to HRTEM, the crystallite size varies from as high as 50 µm at starting conditions down to about 20 nm after 10 h of milling. As the milling time is increased the particle size was found to be in the order of 20 nm at 10 h, see Fig. 5b. After 10 h of milling, the powder is homogeneous with a Pb/Se atomic ratio of approximately 1:1. The average value of particle size obtained from

Rietveld's refinement was found to be in the order of 23 nm; this value is in the same order of magnitude as those observed by HRTEM.

Conclusions

By using the HEBM process it was found that particle size distributions were decreased from 50 μm down to 20 nm. Simultaneously, to these determinations, the rate of chemical reactions in the powders PbO–Se system under atmospheric conditions were traced. In early stages of milling times one of the species plenty identified was PbSe. Another phase identified was $PbSeO_3$. These findings lead us to believe that reactions (1) to (3) give way to reaction (4). However, $PbSeO_3$ phase is stable only up to 8 h of milling time, beyond that time this phase disappears.
A fact that substantiates previous results is explained in terms of volume cell expansion which manifests itself as a positive enthalpy of formation corresponding to the reaction (4).

Acknowledgement

This work was supported by the Consejo Nacional de Ciencia y Tecnología de México (CONACyT) and SIP-IPN. One of the authors (H. Rojas) would like to thank IPN-PIFI for the financial support. Thanks to Dr. X. Bokhimi, M. en C. M. Aguilar-Franco (IF-UNAM), Dr. F.J. Rodriguez-Gomez (IQ-UNAM) and Dr. V. Garibay-Flebes (IMP) for their technical support.

References

[1] P.R. Soni, in: *Mechanical Alloying Fundamentals and Applications*, Cambridge International Science Publishing, Stevenage, England (2001).

[2] G.B. Pinto, S.R.S. Soares, U.U. Gomes, et al.: Mater. Sci. Forum Vol. 591-593 (2008), p. 131.

[3] D.L. Zhang and T.B. Massalski: J. Mater. Res. Vol.9 (1994), p. 53.

[4] L.D. Schmidt, in: *The Engineering of Chemical Reactions*, edited by K.E. Gubbins, chapter, 9, Oxford University Press, Inc., United States of America, NY (1998).

[5] H. Schmalzried, in: *Solid State Reactions*, translated by A.D. Pelton, Verlag Chemie GmbH, Weinheim/Bergstr., Germany (1971).

[6] A.L.G. Rees, in: *Chemistry of the Defect Solid State*, Wiley, London, NY (1954).

[7] F.A. Costa, W.M. Carvalho, A.G.P. Silva, et al.: Mater. Sci. Forum Vol. 591-593 (2008), p. 154.

[8] H. Rojas-Chávez, S. Díaz-de la Torre, et al.: J. Alloys Compd. 483 (2009), p. 275

[9] J. Rodríguez-Carbajal: *FullProf*; Laboratoire Leon Brilloin (CEA-CNRS), Saclay, France 2006.

[10] M. Kara, K. Kurki-Suonio: Acta Crystallogr. A Vol. 37 (1981), p. 201.

[11] I. Barin, in: *Thermochemical data of pure substances*, 3rd Ed., VCH, Weinheim, NY (1995).

[12] *Chemical Thermodynamics of Selenium*, edited by A. Olin, B. Noläng, L-O. Öhman, and E. Rosén, volume 7 of Chemical Thermodynamics, chapter, 5, Elsevier B.V., Amsterdam (2005).

[13] R.A. Swalin, in: *Thermodynamics of Solids*, John Wiley & Sons, Inc., USA (1967).

[14] Q. Li, Y. Ding, M. Shao, et al.: Mater Res Bull Vol. 38 (2003), p. 539

Materials Science Forum Vol. 644 (2010) pp 75-78

doi:10.4028/www.scientific.net/MSF.644.75

Study of hafnium (IV) oxide nanoparticles synthesized by polymerized complex and polymer precursor derived sol-gel methods

R. Ramos-González[1,a], L.A. Garcia-Cerda[1,2,b], H.N. Alshareef[2,3,c], B.E. Gnade[2,d] and M.A. Quevedo-Lopez[2,e]

[1]Centro de Investigación en Química Aplicada. Blvd. Enrique Reyna Hermosillo140 C.P.25250 Saltillo, Coahuila, México.

[2]Department of Materials Science and Engineering, University of Texas at Dallas, Richardson, Texas, USA.

[3]Department of Materials Science, King Abdullah University of Science and Technology Jeddah, Saudi Arabia.

[a]rodolph_rg@yahoo.com.mx, [b]lagarcia@ciqa.mx, [c]husam.alshareef@kaust.edu.sa, [d]gnade@utdallas.edu, [e]mquevedo@utdallas.edu

Keywords: Sol-gel, polymerized complex method, polymer precursor method, hafnium oxide, nanoparticles

Abstract. This work reports the preparation and characterization of hafnium (IV) oxide (HfO_2) nanoparticles grown by derived sol-gel routes that involves the formation of an organic polymeric network. A comparison between polymerized complex (PC) and polymer precursor (PP) methods is presented. For the PC method, citric acid (CA) and ethylene glycol (EG) are used as the chelating and polymerizable reagents, respectively. In the case of PP method, poly(acrylic acid) (PAA) is used as the chelating reagent. In both cases, different precursor gels were prepared and the hafnium (IV) chloride ($HfCl_4$) molar ratio was varied from 0.1 to 1.0 for the PC method and from 0.05 to 0.5 for the PP method. In order to obtain the nanoparticles, the precursors were heat treated at 500 and 800 °C. The thermal characterization of thc precursor gels was carried out by thermogravimetric analysis (TGA) and the structural and morphological characterization by X-ray diffraction (XRD) and transmission electron microscopy (TEM). The XRD patterns of the samples obtained by both methods shows the formation of HfO_2 at 500 °C with monoclinic crystalline phase. The PC method exhibited also the cubic phase. Finally, the HfO_2 nanoparticles size (4 to 11 nm) was determined by TEM and XRD patterns.

Introduction

Several different routes have been reported for the production of metal oxides nanoparticles. Among those techniques sol-gel processing is known to have several advantages, including low temperature processing, good composition control and the ability to develop processes for large areas applications. There are two sol-gel routes that involve the formation of an organic polymeric network. The first is an *in-situ* polymerization route where the gel network is produced by polymerizing the monomers. This is normally achieved in the presence of metal ions. The process is known as polymerized complex (PC) method [1]. An example of this is the Pechini method [2]. The second route involves the preparation of a viscous solution system containing metal ions, polymers and a suitable solvent, and the process is called polymer precursor (PP) method [3].

Interest in the MO_2 oxides, where M = Hf, Ti and Zr, is stimulated by their wide range of industrial applications [4]. Hafnium (IV) oxide (HfO_2) has been extensively studied as a potential alternative to silicon dioxide due to its high dielectric constant and relatively high thermal stability [5]. HfO_2 is also regarded as one of the most commonly used high index coating materials for optical components. This is mostly due to its superior properties such as thermal stability, relatively wide band gap energy, and transparent spectral range from the infrared to the ultraviolet [6].

In the present work, HfO_2 nanoparticles have been prepared by two sol-gel methods: polymerized complex and polymer precursor. The structural and morphological properties of obtained nanoparticles were characterized by X-ray diffraction (XRD) and transmission electron microscopy (TEM).

Experimental Details

For the polymerized complex (PC) method, three different samples were prepared varying the molar ratio (0.1, 0.5 and 1.0) of hafnium (IV) chloride ($HfCl_4$). Firstly, 1 mol of citric acid (CA) was dissolved in distilled water, followed by the addition of the corresponding amount of $HfCl_4$. After complete dissolution 4 mol of ethylene glycol (EG) was added. This process is carried out at room temperature. Next, the solution was heated at ~90 °C under stirring conditions to remove excess water and to accelerate polyesterification reactions. This treatment produced a viscous material that became a semi-transparent glassy resin upon cooling. For the samples prepared by the PP method, the precursor solution was prepared using the procedure described above (PC method), with the exception that CA was replaced by polyacrylic acid (PAA) and without addition of EG. The molar ratio of $HfCl_4$ was varied from 0.05 to 0.5. In order to obtain the HfO_2 nanoparticles, the resin prepared with these methods were heat-treated in a furnace at 500 and 800 °C in air for 2 h.

After synthesis, the nanoparticles were characterized by thermal analyses (TGA), X-ray diffraction (XRD), and transmission electron microscopy (TEM). A TA Instruments Q500 Thermogravimetric Analyzer was used for the resins analyses. Measurement was carried out in air from 25 to 800 °C with a heating temperature rate of 10° C/min. The XRD measurements were obtained using a Siemens D-5000 diffractometer. A CuKα radiation tube was operated at 35 kV and 25 mA. The X-ray powder diffraction patterns were taken in the range of 20-70°, in steps of 0.02°. The particle size, *D*, was calculated for the samples using Scherrer's equation [7],

$$D = \frac{k}{\beta \, c\iota}, \qquad (1)$$

where k is the shape coefficient for the reciprocal lattice point (k = 0.9), λ is the wavelength of incident radiation (CuKα = 0.154 nm), β is the full width at half maximum (FWHM) of the peak and θ is the Bragg angle. Transmission electron microscopy was carried out using a JEOL 2100 HRTEM.

Results and Discussion

Fig.1a shows the TGA curves of the resins prepared by the PC method with different $HfCl_4$ molar ratios. There are three weight losses for each of the samples. The first occurs between 50 and 200 °C which and is due to water present in the resins. The second happens from 200 to 350 °C and is due to the decomposition of un-reacted organic material. Finally, the last weight loss happens between 350 and 600 °C and this is due to combustion of reacted organic compounds [8]. For temperature higher than 600 °C, there is no weight loss in the TGA curves. The percentage of weight loss is related to the initial amount of $HfCl_4$ present in the sample. For the sample with 0.1 mol there is a 99.8% loss, the 0.5 mol shows a 98.4% loss and the 1 mol showed a loss of 96.2%. Fig. 1b shows the TGA curves of the resins obtained by the PP method. Three weight losses are also present for the samples. The first mass loss (50-200 °C) corresponds to elimination of water. The second, the weight loss in the temperature range from 200 to 350 °C, is due to pyrolysis of the PAA [9]. The last, after 400 °C, is due to degradation in form, mainly of CO_2, CO and other gases. 600 °C onwards, significant changes in weight are not observed.

Fig. 2 shows the XRD curves of the HfO_2 synthesized by the polymerized complex and polymer precursor methods using different molar ratios of $HfCl_4$ and thermally treated at 500 and 800 °C.

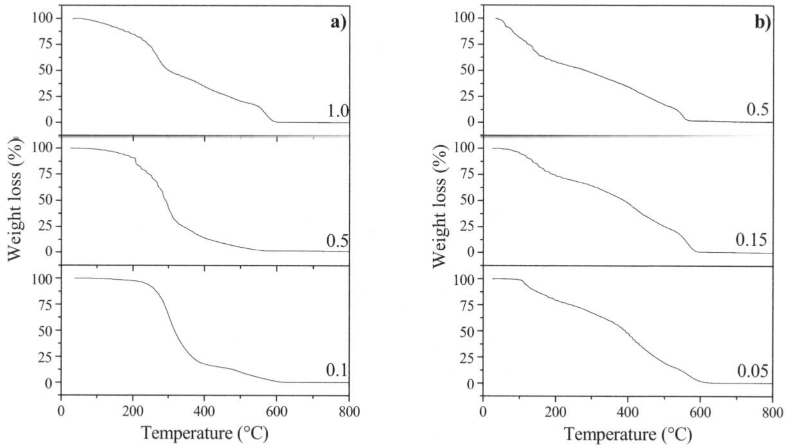

Fig.1. TGA curves of the resins prepared with different $HfCl_4$ molar ratios by the PC (a) and PP (b) derived sol-gel methods.

For reference, the monoclinic and cubic crystallographic patterns of the HfO_2 [10, 11] are also shown. The sample with 0.1 mol of $HfCl_4$ synthesized by PC method heat-treated at 500 and 800 °C (Fig. 2a) show only a single phase (monoclinic) and the samples prepared with 0.5 and 1 mol of $HfCl_4$ shows the monoclinic and cubic phases.

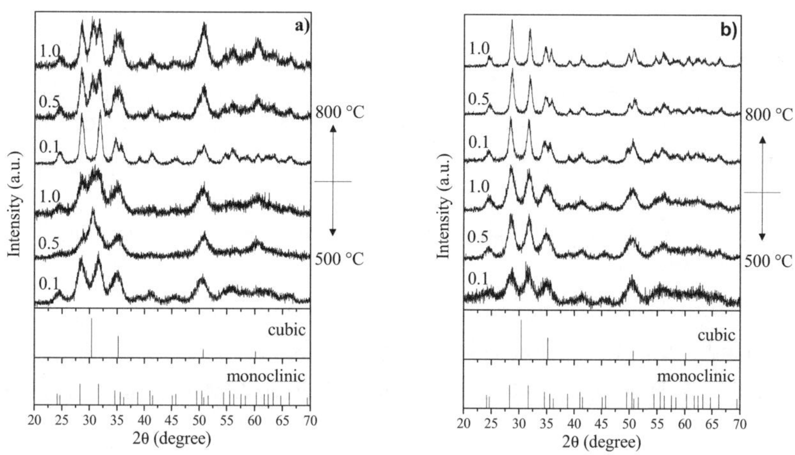

Fig.2. XRD patterns of the HfO_2 synthesized by the PC (a) and PP (b) methods with different $HfCl_4$ molar ratio. Heat-treated at 500 and 800 °C by 2h.

By increasing the concentration of $HfCl_4$, a mixture of phases is present regardless of the heat treatment temperature. All the samples obtained by the PP method (Fig. 2b) shows the characteristic peaks of the monoclinic phase. Particle sizes were calculated using Scherre's equation [7]. The particle size of the samples synthesized by the PC method was from 3 to 10 nm, and by the PP method was between 10 and 18 nm.

The particle size of the sample synthesized by the PC method with 0.1 mol of $HfCl_4$ and heat-treated at 500 °C was also evaluated by TEM (Fig. 3). The majority of the particles have sizes in the range of 8 to 10 nm which is in accordance with those obtained using Scherrer's equation. Agglomerated particles with an irregular shape can be observed.

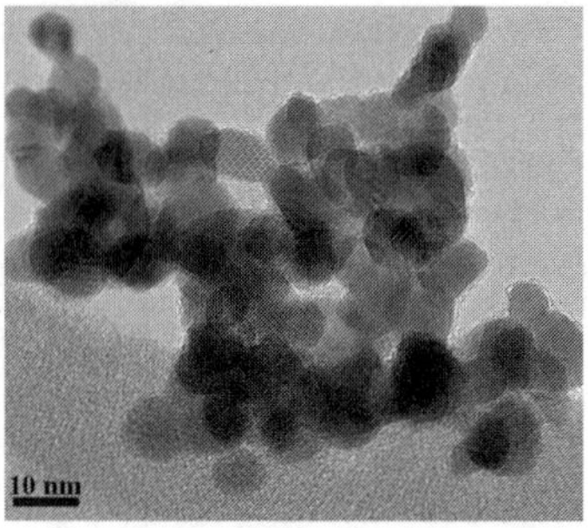

Fig.3. TEM image of the sample synthesized by the PC method with 0.1 mol of $HfCl_4$. Heat-treated at 500 °C by 2 h.

Conclusions

We reported a comparison between polymerized complex (PC) and polymer precursor (PP) derived sol-gel methods for the synthesis of HfO_2 nanoparticles. The particles can be obtained at temperatures as low as 500 °C with average size 10 nm. Increasing the $HfCl_4$ concentration, cubic and monoclinic phases of HfO_2 nanoparticles is obtained by PC method. The nanoparticles prepared by PP method only a monoclinic phase is obtained.

Acknowledgements

The authors would thank CONACyT for the economical support for R. Ramos-González (scholarship 170601) and L.A. García-Cerda (sabbatical leave) in the University of Texas at Dallas.

References

[1] M. Kakihana, M. Yoshimura, H. Mazaki and H. Yasuoka: Rep. Res. Lab. Eng. Mater., Tokyo Inst. Tech. Vol. 17 (1992), p. 63.

[2] M.P. Pechini, U.S. Patent 3,30,697. (1967)

[3] M. Kakihana: J. Sol-Gel Sci. Tech. Vol. 6 (1996), p. 7

[4] S.M. Woodley, S. Hamad, J.A. Mejías and C.R.A. Catlow: J. Mater. Chem. Vol. 16 (2006) p. 1927

[5] Xinye Liu, S. Ramanathan, A. Longdergan, A. Srivastava, E. Lee, T.E. Seidel, J.T. Barton, D. Pang and R.G. Gordon: J. Electrochem. Soc. Vol. 152 (2005), p. G213

[6] J.M. Khoshman, A. Khan and M.E. Kordesch: Surf. Coat. Technol. Vol. 202 (2008) p. 2500

[7] B.D. Cullity: *Elements of X-ray Diffraction* (Addison-Wesley, Reading, MA. 1978).

[8] H. Guo and Y. Qiao: Appl. Surf. Sci. Vol. 254 (2008) p. 1961

[9] S. Yang-Kook, O. In-Hwan and H. Seong-Ahn: J. Mater. Sci. Vol. 31 (1996) p. 3517

[10] JCPDS, International Center for Powder Diffraction Data: Swarthmore, PA, 1989; Card No. 074-1506. J. Am. Ceram. Soc. Vol. 53 (1970) p. 126

[11] JCPDS, International Center for Powder Diffraction Data: Swarthmore, PA, 1989; Card No. 053-0550.

Materials Science Forum Vol. 644 (2010) pp 79-84
© (2010) Trans Tech Publications, Switzerland
doi:10.4028/www.scientific.net/MSF.644.79

Preparation of nano-ceramics via aqueous sol-gel method modified with surfactants: an overview

Yibran Perera[1,a]

[1]Centro de Investigación en Química Aplicada (CIQA), Blv. Ing. Enrique Reyna Hermosillo, No. 140, C.P. 25100, Saltillo, Coahuila, México.

[a]yperera@ciqa.mx

Keywords: sol-gel, surfactants, silica, alumina, di- and tri-calcium silicate, oil-well cement.

Abstract. The present paper shows an overview related to the preparation and characterization of different nano-materials such as: silica, alumina, di- and tri-calcium silicates and a composite in the ternary CaO-SiO_2-Al_2O_3 system. These nano-materials are related to the construction area so some basic tests are carried out to evaluate their cementitious and/or pozzolanic behavior. The nano-ceramics have been synthesized via hydrolysis and condensation of aqueous sol-gel method modified with surfactants using different precursors such as: tetraethyl orthosilicate (TEOS), calcium nitrate tetrahydrate ($Ca(NO_3)_2.4H_2O$), sodium silicate ($Na_2O.SiO_2$), aluminum nitrate nonahydrate ($Al(NO_3)_3{\bullet}9H_2O$) and an industrial aluminum hydroxide (gibbsite). The synthesis parameters such as: pH, temperature, kind of surfactants, etc., are reported. To characterize the structural, chemical composition, particle size and morphology of the nanoparticles, different techniques such as: Fourier transform infrared (FTIR), thermal gravimetric analysis (TGA), differential scanning calorimetry (DSC), X-ray diffraction (XRD), field emission electron microscopy (FESEM) and high-resolution transmission electron microscopy (HR-TEM) are used.

Introduction

Silica and its related materials have become increasingly used in diverse industries such as: electronic, chemical, pharmaceutical, construction, medical and with the recent nanotechnological improvement in the ceramic materials area, also the oil industry is showing an increased demand for well-defined silica and silica-based materials [1,2]. A variety of nano-ceramics can be synthesized chemically by the universal sol-gel method giving excellent opportunities for structural modification of inorganic materials, e.g. the production of silica by a direct templating mechanism has been used in the last years by many research groups of colloidal sciences [1,2,3]. In this way, the sol-gel method modified with surfactants has been successfully used in the production of different nanomaterials; specially, for the production of nanostructured materials with a variety of mesostructures that are attained by the cooperative self-assembly of organic surfactants and inorganic species [1,3,4]. On the other hand, at present, the construction area in the entire world is going through a new route of changes related to the nanomaterials, specifically for systems such as ordinary Portland cement, concrete, mortars and also to produce new oil-well cement slurry formulations using different nanoparticles [5,6]. In this way, a range of nanocements are available and produced to solve specific industrial problems. Also, recent research has been focusing on the investigation and production of new construction nanomaterials giving the chance to study the unexplored cement and the cement-based materials areas [7,8,9]. The aim of this paper is to report the advance in the sol-gel chemistry combined with surfactant science, opening the possibility to generate nano-ceramics with potential applications in the construction materials. In adittion, it is important to mention the excellent synergy generated between the production of nano-ceramics and the construction area, using the nanoscience and the nanotechnology as an innovator bridge that will offer improvements in the next future to the world construction systems.

Experimental methodology

Set-up: Aqueous sol-gel method modified with surfactants using organic and inorganic precursors was carried out to synthesize different nano-ceramic materials. For all the cases the gels obtained were transparent. The parameters employed in these experiments are shown in Table 1.

Table 1. Experimental parameters used for the production of the nano-ceramics.

Ceramics	Precursors	pH	Acid	Sol Temp. [°C]	Surfactants	Particles size [nm]	Ref.
SiO_2	Sodium Silicate & NH_4OH	2-4	HCl	<5 & room temp	Triton-X100	< 20	10,11
	TEOS, Ethanol and NH_4OH	3-5		50	CTABr & Pluronic F127	5 - 30	1,3
$2CaO\cdot SiO_2$	TEOS, Ethanol, NH_4OH & $Ca(NO_3)_2.4H_2O$	3-5	HNO_3	40-60	Pluronic P130	< 15	2,14
$3CaO\cdot SiO_2$							
Al_2O_3	Gibbsite, Ethanol & NH_4OH	2-5	HCl + HNO_3	40-80	Triton-X100	< 20	12
Ca-Si-Al	$Al(NO_3)_3$•$9H_2O$, TEOS, Etanol, NH_4OH & $Ca(NO_3)_2.4H_2O$	1-4	HNO_3	25-70	Pluronic P104	Grains ≈ 50	13

Preparation details: The different nano-ceramics are produced by the adequate and controlled parameters shown in the table 1; the most important details for each system are shown as following:

a) Nano-Silica:

- **Sodium Silicate as precursor.** The sodium silicate is mixed with water and stirred for few minutes (pH ~ 13); this basic solution is cooled at temperatures <5°C, then an acid solution is added until a pH ~ 2 is obtained. At this pH the monosilic acid is present in the solution and enclosed into the non-ionic surfactant micelles system. Finally, the sol-gel process is performed.
- **TEOS as precursor.** The TEOS is mixed with Ethanol and stirred for one hour. A double surfactant system was used as follows: the CTABr cationic aqueous solution is added to form the first self-assembly system with the inorganic silicon anionic species at pH ~ 3; Then, the colloidal system is encapsulated using the non-ionic Pluronic F127 micelles as a secondary surfactant system.

b) Nano-Calcium Silicates. The $CaO:SiO_2$ molar ratio in the starting mixture were kept to 2:1 and 3:1 in order to obtain $2CaO.SiO_2$ and $3CaO.SiO_2$ respectively. The nanoparticles in all cases were obtained using non-ionic surfactant confining systems. The ceramic phase transformations of these systems are studied using an adequate thermal treatment up to 900 and 1400°C, respectively.

c) Nano-Alumina. Industrial gibbsite is treated with a mixed acid solution of HNO_3/HCl in a microwave system during 15 min to obtain a concentrated liquor sol. The sol is diluted and mixed with the non-ionic surfactant solution until the final transformation from sol to gel is reached. The products are heated up to 1150 °C to study the phase transformations that occur to obtain the α-Al_2O_3 phase.

d) Nanostructured material in the ternary CaO-SiO_2-Al_2O_3 system. The mixture of TEOS and Ethanol was stirred for one hour to prepare a first solution with the silicon species. Also two different solutions of aluminum and calcium salts in distilled water and/or Ethanol were prepared. These solutions are mixed with the non-ionic aqueous surfactant solution. The products are heated up to 1350 °C to study the phase transformations of this system.

Characterization techniques: A Fourier transform infrared spectrometer was utilized for Fourier transform infrared (FTIR) analysis with KBr pellets for all the evaluated samples. These studies are focused on the region between: 450-4000 cm^{-1}. For all samples, the gels were dried at temperature between 80-100°C. Thermal gravimetric analyses (TGA) and differential scanning calorimetry

(DSC) analyses were used to fix the thermal treatment required for the phase transformations in each system. The diverse ceramic powders produced at different temperatures were analyzed by X-ray diffraction (XRD). Finally, the samples were studied by high-resolution transmission electron microscopy (HR-TEM). In addition, the nano-SiO_2 particles were used as pozzolanic material to produce Portland cement slurries and the compressive strength of the samples cured at 7th days was measured. On the other hand, the *in situ* nano-calcium silicates hydration process (both cases) was observed by field emission scanning electron microscopy (FESEM).

Results and discussion

Gels analysis

In this step, the aim was the production of transparent gels. The evaporation of exceeding solvent (water/ethanol), the phase (solution/gel) separations and the sineresis of the gels were evident for all studied cases. To understand the chemical process related to the production of chemical species developed in each system different analysis were carried out. In this way, the FTIR analysis indicated that water is identified in all cases because at these temperatures the materials have a high amount of hydroxide phases. The spectra of nano-SiO_2 and nano-calcium silicates show vibrations related to the chemical bonding like Si-O-Si, Si-OH and Si-O-Ca. For the specific case of the nanostructured CaO-SiO_2-Al_2O_3 material, the obtained spectrum also shows additional signals such as: Si-O-Al and Al-OH-Al. Finally, the gel precursor of Alumina reported some vibrations from bondings like Al-OH and Al-OH-Al. The vibration bands at their range of frequencies for each case are shown in Table 2.

Table 2. Different vibration bands from each precursor gels obtained by FTIR spectral.

<table>
<tr><th rowspan="2">GEL</th><th colspan="9">Frequency [cm^{-1}]</th></tr>
<tr><th>450-470</th><th>550</th><th>580-700</th><th>835</th><th>950-1000</th><th>1050-1250</th><th>1650</th><th>2830-3000</th><th>3000-3750</th></tr>
<tr><td>SiO_2*</td><td rowspan="3">Si-O-Si</td><td rowspan="2">-</td><td rowspan="3">-</td><td rowspan="3">Si-O-Si</td><td>Si-(OH)</td><td rowspan="3">Si-O-Si</td><td rowspan="4">H_2O</td><td rowspan="4">C-H</td><td rowspan="2">O-H, Si-OH; H_2O</td></tr>
<tr><td>$2CaO\cdot SiO_2$ $3CaO\cdot SiO_2$</td><td>Si-O-Ca</td></tr>
<tr><td>Ca-Si-Al</td><td>Si-O-Al</td><td>Si-O-Ca; Al-OH-Al</td><td>Ca-OH H_2O</td></tr>
<tr><td>Al_2O_3**</td><td>Al-OH</td><td colspan="2">Al-OH-Al</td><td>-</td><td colspan="2">Al-OH-Al</td><td>Al-OH; H_2O</td></tr>
</table>

*Nano-SiO_2 from TEOS as precursor. For the case using sodium silicate the vibration at frequency at 2830-3000 cm^{-1} is not detected.
**Diverse Al-OH vibrations are detected from this system and associate to the aluminum hydroxide gel precursor.

Phase transformations of nano-ceramics

Gels were thermally treated to generate the final nano-ceramics. The temperature selection was according to the ones results obtained by TGA and DSC analysis from each studied system. The XRD results are summarized in Table 3. Nano-calcium silicates show a production of different final ceramic phases such as: $2CaO\cdot SiO_2$ (α, α' and β) and $3CaO\cdot SiO_2$ (monolitic); Similarly, the CaO-SiO_2-Al_2O_3 nanostructured material shows a series of phase transformations that occur to generate a mixture of final phases such as: $2CaO\cdot SiO_2$ (α, α' and β), $3CaO\cdot SiO_2$ (monolitic) and $3CaO\cdot Al_2O_3$. These phases are important because all of them have different levels of hydraullic properties.

On the other hand, nano-SiO_2 thermally treated up to 600°C does not show any crystallization phase. Finally, for the specific case of the aluminium hydroxide crystalline transformations into nano-Al_2O_3, the XRD analysis shows the production of α-alumina at 1200°C. The reported route was as follows: $Pseudo-Bohemite \rightarrow \gamma - Al_2O_3 \rightarrow \alpha - Al_2O_3$.

Table 3. Nano-ceramic phases obtained from the XRD analysis.

<table>
<tr><th rowspan="2">Temperature [°C]</th><th colspan="5">Ceramic Crystalline Phases</th></tr>
<tr><th>SiO_2*</th><th>$2CaO·SiO_2$</th><th>$3CaO·SiO_2$</th><th>Al_2O_3</th><th>$CaO-SiO_2-Al_2O_3$</th></tr>
<tr><td>80-100</td><td colspan="5">Amorphous</td></tr>
<tr><td>600</td><td>Amorphous</td><td colspan="2">-</td><td rowspan="3">$\gamma-Al_2O_3$</td><td rowspan="2">$\alpha-2CaO·SiO_2$, $Ca(OH)_2$, CaO & $\gamma-Al_2O_3$</td></tr>
<tr><td>700</td><td colspan="3">-</td></tr>
<tr><td>800-900</td><td rowspan="3">-</td><td>$\alpha-2CaO·SiO_2$; $\alpha'-2CaO·SiO_2$; & $\beta-2CaO·SiO_2$</td><td rowspan="2">-</td><td>-</td></tr>
<tr><td>1200</td><td rowspan="2">-</td><td>$\alpha - Al_2O_3$</td><td>$\beta-2CaO·SiO_2$; $\gamma-2CaO·SiO_2$; $3CaO·Al_2O_3$ & CaO</td></tr>
<tr><td>1350 - 1400</td><td>$3CaO·SiO_2$ (Monolitic) & CaO</td><td>-</td><td>$3CaO·SiO_2$ (Monolitic); $\beta-2CaO·SiO_2$; $3CaO·Al_2O_3$ & CaO</td></tr>
</table>

*Using both precursors: TEOS and Sodium Silicate.

Morphological characterization of nano-ceramics

The morphologies from each nano-ceramic are studied by HR-TEM and are shown at Fig 1. The nanoparticles have a spherical geometry with the particle sizes showed in Table 1. The nanostructured material $CaO-SiO_2-Al_2O_3$ system and also the nano-Al_2O_3 particles showed aggregations. This fact is due to the particle sinterization/agregation process related to the high temperatures used during the thermal treatment.

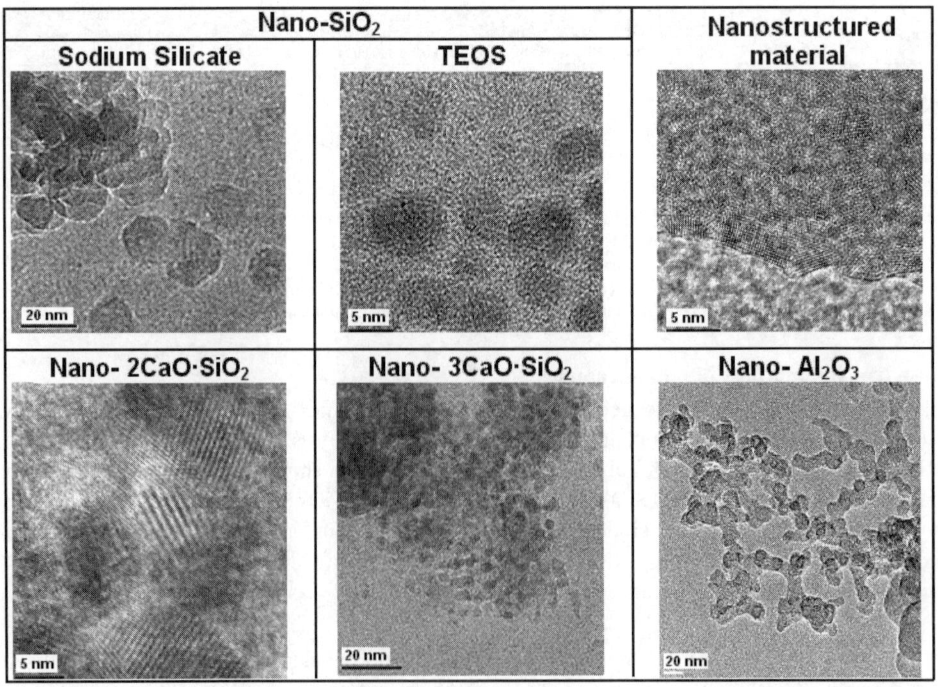

Figure 1. HR-TEM images. The nano-SiO_2 [1,3,10,11] and nano-$2CaO-SiO_2$ / nano-$3CaO-SiO_2$ [2,14] systems show spherical morphologies. The nano-Al_2O_3 system indicates an agglomeration of spherical nanoparticles [12]. Finally, the nanostructured material is constituted by nano-polycrystalline grains [13].

Some important properties related to nano-ceramics

To know about the cementitious and/or pozzolanic properties developed from these kinds of nano-ceramics some evaluations were carried out and summarized at Fig. 2. Specifically, Fig. 2a shows how the ordinary Portland cement can increase some mechanical properties, e.g. compressive strength when it is mixed with nano-SiO_2 (1 and 2 %wt.). The samples analyzed at the 7th day show an increase in their compressive strength of 54% (with 1% of nano-SiO_2) up to a value of 147% when 2% of nano-SiO_2 is used. The increases in the compressive strength are related to the increases in the amount of C-S-H gel generated when the nano-SiO_2 are introduced into the cement hydration process. According to Ye Qing [15], the nano-SiO_2 has many unsaturated bonds $\equiv Si{-}O{-}$ and $\equiv Si{-}$ in the surface. The reactions process between SiO_2 and the $Ca(OH)_2$ (generated during the cement hydration) may be as follows:

$$\equiv Si - O - + H - OH \rightarrow \equiv Si - OH \quad \text{(reacts quickly)} \quad (1)$$

$$\equiv Si - + OH \rightarrow \equiv Si - OH \quad \text{(reacts quickly)} \quad (2)$$

$$\equiv Si - OH + Ca(OH)_2 \rightarrow C - S - H_{(gel)} \quad (3)$$

On the other hand, the Fig. 2b shows the field-emission scanning electron microscopy (FESEM) image obtained in situ during the chemical interaction of the nano-calcium silicates with the atmospheric water. In this image it is possible to observe how an aggregate of nanoparticles can be directly transformed to cement products such as: C-S-H gel and calcium hydroxide ($Ca(OH)_2$).

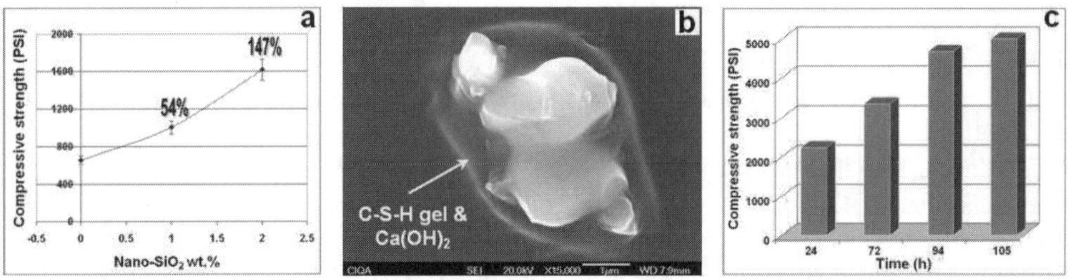

Figure 2. Results indicating the pozzolanic/hydraulic properties of the nano-ceramics. a) Nano-SiO_2 used as additive into conventional Portland cement slurries [8]. b) Transformation in-situ of nano-calcium silicates into cement products (FESEM image) [2,14]. c) Increase of the compressive strength from a "G" oil-well cement system with nano-SiO_2 [6].

Finally, the petroleum industries, specifically the oil and gas exploration and production areas, are interested in the use of new nano-ceramics that could be employed to improve, for instance, mechanical properties of the oil-well cement systems [1,2,6,14]. Fig. 2c shows how the compressive strength of the G oil-well cement system increases with the use of a commercial product based on nano-SiO_2 named NS-33C from Ulmen Company. The G oil-well cement slurry was produced with a density of 16.8 LPG and a mixture of additives such as: 0.01 gps antifoam, 0.15% retardant, 0.3% fluid loss controller, 20% silica and 0.25 gps NS-33C. The compressive strength of this cement shows higher values at minor time compared to a conventional G oil-well cement system [6].

Conclusions

Diverse nano-ceramics with different particle sizes were produced by the sol-gel method modified with surfactants. The surfactant micelles could encapsulate and disperse the different nanoparticles during the synthesis process into the aqueous solutions. The fine ceramics are produced after an adequate thermal treatment. This production methodology can be inexpensive if the precursors are from industrial processes such as the production of nano-Al_2O_3 synthesized/reported in this paper. Specifically, the nano-SiO_2 particles stabilized/dispersed in aqueous solution with particle size under 30 nm showed a high pozzolanic properties when they are incorporated in a relation of 2 %wt to the cement into the ordinary Portland cement slurries. Also an increase of 147% in compressive strength is observed in samples cured during 7 days. Actually, the strength improvement of the Portland cement using nano-SiO_2 is being applied in different construction activities; e.g., petroleum corporations would generate cementing well systems with better mechanical properties in general terms. It is based on the nano-SiO_2 pozzolanic reactions with the cement slurries resulting in the consumption of $Ca(OH)_2$ and formation of an additional C-S-H gel that is the principal product to supply the increase of the compressive strength of any cement system. On the other hand, the CaO-SiO_2-Al_2O_3 nanostructured material, constituted by cement-phase grains with size of 50 nm, showed similar cementitious properties than the nano-calcium silicates, which have higher reactivity with the atmospheric water, generating C-S-H gel as principal hydration product and $Ca(OH)_2$ as byproduct. Finally, the cementitious/pozzolanic properties of these kinds of ceramics offer a new route of applications in diverse industries related to the construction area over the world.

Literature References

[1] Y. Perera, J. Cano y S. Martínez; *Síntesis de nanopartículas de sílice vía sol-gel para aplicaciones en cementación de pozos petroleros*, XII congreso Venezolano de Microscopia, Pto. Ordaz, Venezuela, (2006), abstract and poster.

[2] Y. Perera, J. Cano, S. Martínez, G. Quercia; *Characterization of nano-cement phases by Field emission scanning electron microscopy (FESEM)*, abstract and poster at 9th CIASEM-Acta Microscópica. Vol.16, (2007), No.1-2.

[3] J. Cano; *Síntesis y caracterización de nano-partículas de sílice empleando surfactantes vía sol-gel*, B.S. Thesis; Universidad de los Andes (ULA). Mérida, Venezuela (2007).

[4] D. W. Schaefer: MRS Bull. Vol.14 (1994).

[5] J. H. Johnston, A. J. McFarlane, T. Borrmann and J. Moraes; Current Applied Physics 4 (2004) pp. 411-414.

[6] Information on http://www.ulmen.cl

[7] P. Balaguru and K. Chong; Proceedings of ACI session on: *Nanotechnology of Concrete: Recent Developments and Future Perspectives*; Denver, USA (2006).

[8] P. Taylor, K. Rajan, B. Birgisson and T. Cackler; *Workshop on Nanotechnology for cement and concrete* (2007). Information on: http://www.intrans.iastate.edu/cncs/nanotech-wkshprpt.pdf

[9] Z. Bittmar, P.J.M. Bartos, J. Nemecek, V. Smilauer, and J. Zeman; *Nanotechnology in Construction 3*, Proceedings of the NICOM3, Springer, (2009).

[10] J. Uzcátegui; *Estudio de la química sol-gel modificada con surfactante no-iónico para la obtención de nano-SiO_2 a partir de silicato de sodio*, 1st seminar related to the B.S. Thesis. Universidad de los Andes (ULA). Mérida, Venezuela (2008).

[11] K. Agafonow; *Estudio de la interacción de un nanoaditivo sobre la técnica de estabilización y solidificación de cortes de perforación (ESCP),* B.S. Thesis, Universidad Simón Bolívar (USB). Caracas, Venezuela (2009).

[12] M. Morales; *Síntesis y caracterización de nano-Al_2O_3 empleando surfactantes,* 1st and 2nd seminars related to the B.S. Thesis. Universidad de los Andes (ULA). Mérida, Venezuela (2009).

[13] A. Hurtado; *Preparación de partículas manométricas del sistema ternario CaO-SiO_2-Al_2O_3 vía sol-gel modificado con surfactante no-ionico*, B.S. Thesis; Universidad Central de Venezuela (UCV). Caracas, Venezuela (2008).

[14] Y. Perera; *Micro- y Nano-cerámicas nacionales para operaciones de cementación de pozos de Hidrocarburos*, Encuentro Académico Industrial JIFI-EAI 2008. abstract and special oral presentation. Caracas, Venezuela (2008).

[15] Y. Qing, Z. Zenan, K. Deyu, Ch. Rongshen; Influence of nano-SiO2 addition on properties of hardened cement paste as compared with silica fume, Const. and building Materials, 21, 539-545, (2005).

Materials Science Forum Vol. 644 (2010) pp 85-90
© (2010) Trans Tech Publications, Switzerland
doi:10.4028/www.scientific.net/MSF.644.85

On the Influence of Silver Nanoparticles Size in the Electrical Conductivity of PEDOT:PSS

Rebeca G. Melendez[1], Karla. J. Moreno[1,a], Ivana Moggio[1], Eduardo Arias[1], Arturo Ponce[1], Irantzu Llanera[2], Sergio E. Moya[2]

[1]Centro de Investigación en Química Aplicada, Boulevard Enrique Reyna 140, 25253, Saltillo, Mexico

[2]CIC biomaGUNE Centro de Investigación Cooperativa en Biomateriales, Parque Tecnológico de San Sebastián, 20009, San Sebastián (Gipuzkoa) España

[a]karlamoreno@itc.mx

Abstract. In this paper, we studied the influence of the silver nanoparticles size on the electrical conductivity of PEDOT:PSS in Ag(PEDOT:PSS) films. The silver nanoparticles were synthesized in presence of PEDOT:PSS by varying the molar ratio between $AgNO_3$ and the reducing agent ($NaBH_4$). Both the particle size determined by TEM and the plasmon band obtained by UV-Vis spectroscopy were found to be strongly dependent on the reducing agent concentration. The electrical conductivity increases inversely with the concentration of reducing agent from 5.24×10^{-4} up to 1.63 S/cm; three orders of magnitude higher than pristine PEDOT:PSS.

Introduction

Poly(3,4-ethylenedioxythiophene)-poly-(styrenesulfonate) (PEDOT:PSS) is one of the most promising candidate for application in different electronic devices, mainly due to its good electrical conductivity [1]. PEDOT is commercially available in a non stechiometric complex with PSS which permits its dispersion in water to obtain thin films as PEDOT alone cannot be easily processed. However, this complexation is accompanied by a decreasing in the electrical conductivity of PEDOT [2]. Recently, much effort has been focused on increasing the conductivity of PEDOT:PSS by adding liquid or solid organic compounds, such as methyl sulfoxide (DMSO), glycerol or sorbitol to the PEDOT:PSS aqueous suspension [3-5]. However, the mechanism of conductivity enhancement in the PEDOT:PSS is still controversial. For example, X-ray photoelectron spectroscopy (XPS) suggested that when adding appropriated amount of sorbitol to PEDOT:PSS, the insulating PSS chains are rinsed way from the surface of the film, leaving a high conductive thin layer composed of only PEDOT chains [6]. A later work argues that when adding organic compounds, the conductivity enhancement in the PEDOT:PSS is due to the tautomerism of thiophene moieties of the PEDOT chain from benzoid to the quinoid form [7]. On the other hand, a different strategy to improve the electrical conductivity of a conjugated polymer is through the incorporation of metal nanoparticles into the conducting polymer [8-10]. Recently, we have reported the *in situ* functionalisation of silver nanoparticles with the conjugated polymer PEDOT:PSS [11]. We have demonstrated that the electrical conductivity of the silver nanoparticles/polymer composite film, briefly Ag(PEDOT:PSS), is improved with respect to the PEDOT:PSS film, while maintaining the good processability in thin film. In the present work, we survey the influence of different concentrations of the reducing agent on the electrical and optical properties of Ag(PEDOT:PSS) composites. The experimental results provide evidences that with the appropriate amount of $NaBH_4$, the electrical conductivity of PEDOT:PSS can be varied opening the possibility to use the composites in different devices for electronic (in the case of conductive nanocomposites) or optoelectronic (semiconductive nanocomposites) technology.

Experimental

Ag(PEDOT:PSS) composite suspensions were prepared under the same experimental conditions described in ref. [11]. All reagents were purchased from Aldrich and used as received. PEDOT:PSS was a 1.3% wt water dispersion composed of 0.5% wt of PEDOT and 0.8% wt of PSS. Briefly, 5 mL of PEDOT:PSS were added to 10 mL of 2.94 x 10^{-2} M water solution of $AgNO_3$ (99%). 10 mL of different concentrations of reducing agent (see Table 1). The Ag(PEDOT:PSS) suspensions were characterized by UV-Vis spectroscopy on a Shimadzu 2401 PC spectrophotometer. The size and morphology was analyzed by Transmission Electron Microscopy (TEM) (JEOL JEM-2100F, Japan). Particle size distribution was calculated by image analysis over 100 counts. Thin films were prepared by immersion of hydrophilic quartz slides from SPI during 10 min in each Ag(PEDOT:PSS) suspension. The topography of the films was analyzed by Atomic Force Microscopy (AFM) on a Digital Instrument Dimension 3100 in tapping mode at 0.5 Hz. The film thickness was measured using a Dektak 6 profilometer from Veeco with scanning rates of 50 mm/s and applied force of 2 mg. The electrical conductivity study was carried out on thin films prepared as mentioned before by four probe method using an electrometer Kiethley 6517A of high resistance.

Table 1. Concentration of $NaBH_4$ and $AgNO_3$:$NaBH_4$ molar ratios used for the nanocomposites synthesis.

Nanocomposite	[$NaBH_4$], M	Molar ratio, $AgNO_3$:$NaBH_4$
Ag(PEDOT:PSS)-**I**	2.64 x 10^{-3}	10:1
Ag(PEDOT:PSS)-**II**	2.64 x 10^{-2}	1:1
Ag(PEDOT:PSS)-**III**	5.28 x 10^{-2}	1:1.2
Ag(PEDOT:PSS)-**IV**	1.58 x 10^{-1}	1:10

Results and Discussion

The synthesis of silver nanoparticles in the presence of PEDOT:PSS is expected to occur from Ag^+ that is reduced to Ag^0 by $NaBH_4$. The silver atoms grow as clusters forming the nanoparticles. The excess of PSS gives an anionic nature to the PEDOT:PSS complex that promotes the stability of the silver nanoparticles as previously proposed by Kumar et al. [12] for PEDOT:PSS gold nanoparticles. The amount of sodium boron hydride used in a previous work was established as the minimum necessary to observe a color change; which is indicative of the metallic nanoparticles formation, with the idea that the excess of $NaBH_4$ could provoke de-doping of PEDOT:PSS affecting the electrical conductivity of the nanocomposite. In order to better understand the effect of the reducing agent in the composites properties on the electrical conductivity, different concentration of $NaBH_4$ were used in the present work (see Table 1). The concentration was varied in order to have a molar ratio between the metallic salt and the reducing agent passing from 10:1 to 1:10. In the first reaction **I**, there is a strong defect of the reducing agent and not the all quantity of silver +1 is converted to Ag^0. In the second reaction **II**, the molar ratio is close to 1:1 corresponding to pure silver nanoparticles in the PEDOT:PSS matrix. In the third **III** and fourth **IV** reaction, an excess of $NaBH_4$ is used and remains available for eventually further reducing reactions. The amount of PEDOT:PSS was kept constant for the four reactions and in excess with respect to $AgNO_3$, thus part of this amount will form the shell stabilizing layer around the metallic nanocore and part remains as matrix. In all cases, the addition of the reducing agent is accompanied by a colour change of the reaction mixture from blue to brown, confirming the formation of the metallic nanoparticles. Fig. 1 shows TEM image and its statistical distribution of the Ag(PEDOT:PSS)-**II** sample as representative of the silver nanoparticles obtained in the Ag(PEDOT:PSS) composite suspensions .

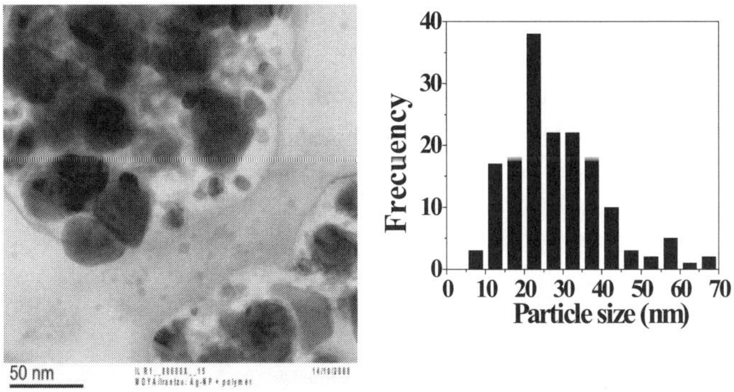

a) b)

Fig. 1. (a)TEM image and (b) particle size distributions of silver nanoparticles in a Ag (PEDOT:PSS)-**II** composite suspension.

In all the reactions, dispersed nanoparticles with elongated spherical morphology and agglomerates can be observed as previously found [11]. The polymer matrix can be also observed as a diffuse halo. By decreasing the reducing agent concentration, the mean size as well as its standard deviation increases (Table 2). Zukoski [13] reports that the silver nanoparticle formation occurs extremely rapidly when the reducing agent is the sodium boron hydride, giving rise to small particles (~2-3 nm), which grow in a second stage to achieve higher sizes (~8-20 nm) between 0.20 to 1 h. In the final stage of the reaction, the boron hydride is consumed by reacting with water causing loss of BH_4^- ions and making that the suspension passes from a reducing to a rather oxidizing medium. This particular environment, favours the particles agglomeration. Our result could be thus explained in terms of the loss of stabilizing species BH_4^- as consequence of decreasing concentration of $NaBH_4$. This were no considered in our previous work [10] where the silver nanoparticles size was smaller than that found in this study as the PEDOT:PSS used from Clevios presented a weight ratio PEDOT:PSS of 1:2.5 against 1:1.6 of the polymer used in this work. We think that the higher content of PSS gave a stronger excess of negative charge limiting the growing of nanoparticles.

Table 2. Summary of properties of Ag(PEDOT:PSS) water suspensions and thin films.

Nanocomposite	Diameter size (nm)	λ_{max}	ε	RMS (nm)	σ (Scm^{-1})
Ag(PEDOT:PSS)-**I**	28.18 ± 12.1	415	5.2 x 10^{-1} gL^{-1}cm^{-1}	2.40	1.63
Ag(PEDOT:PSS)-**II**	18.01 ± 11.7	412	2.3 x 10^{-1} gL^{-1}cm^{-1}	2.38	2.9 x 10^{-1}
Ag(PEDOT:PSS)-**III**	15.01 ± 10.5	400	1.5 x 10^{-1} g L^{-1}cm^{-1}	3.73	2.6 x 10^{-2}
Ag(PEDOT:PSS)-**IV**	11.5 ± 7.7	390	7.0 x 10^{-2} g L^{-1}cm^{-1}	3.04	5.24 x 10^{-4}

The UV-Vis spectra of Ag(PEDOT:PSS) composites and of pristine PEDOT:PSS are reported in Fig. 2. The UV-Vis spectrum of PEDOT:PSS exhibits a strong absorption peak at 240 nm due to the π–π electronic transitions of the aromatic rings and a broad absorption tail between 600 and 1000 nm ascribed to the polaronic band of the conductive quinoid form. When the silver nanoparticles are formed, a strong peak around 400 nm arises, which is characteristic of silver nanospheres surface plasmon [14]. The ε values for this peak increases with increasing the size of the nanoparticles. The polaronic band can be observed only for Ag(PEDOT:PSS)-**I**, where an excess of PEDOT:PSS that is not coordinating the silver nanoparticles derives from the stechiometry of the reaction. For Ag(PEDOT:PSS)-**II** to -**IV** there is no evidence of this band as previously reported for

Ag(PEDOT:PSS) Clevios composite [11] as the plasmon band peak absorption is stronger. The calculated molar extinction coefficient for the plasmon band increases with the size as reported in literature [15-17]. By decreasing the concentration of $NaBH_4$, the plasmon band red shifts consistently with the increase in the nanocore size as shown in previous studies on metallic nanospheres [15-17]. Plasmon band, however, has also been reported to red shift according to increase in the dielectric constant of surrounding medium [18-21]. Theoretical simulation of the UV-Vis spectra is currently in progress to investigate the contribution of the two effects (size and medium) on the optical properties for these nanocomposites.

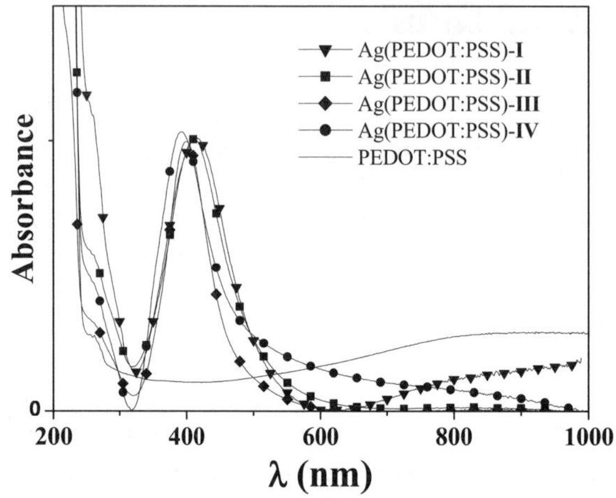

Fig. 2. UV-Vis-NIR spectra of Ag(PEDOT:PSS) composite solutions prepared at different $NaBH_4$ concentrations.

The morphological properties of thin films of Ag(PEDOT:PSS) and PEDOT:PSS were analyzed by AFM and the corresponding topography images are shown in Fig. 3. PEDOT:PSS exhibits a homogeneous and smooth surface, similarly to that reported by Ouyang et al., [7] while the Ag(PEDOT:PSS)-**II** (taken as representative example for the nanocomposite films) image exhibits a granular morphology, where the grains are well distributed and can be likely due to the silver nanoparticles clusters. The same morphology was found for the other three samples analyzed in this work. In the whole, all the films showed low values of the root-mean-square (RMS) roughness in the range of 2.4 to 3.7 nm indicating a very good homogeneity. It was found that the electrical conductivity (σ, Table 2) increases as the silver nanoparticles size increases, enhancing almost five orders of magnitude from Ag(PEDOT:PSS)-**I** to Ag(PEDOT:PSS)-**IV**. It is worth to mention that the electrical conductivity of the pristine PEDOT:PSS films was $5.17 \times 10^{-3}\ Scm^{-1}$. Interestingly, the Ag(PEDOT:PSS)-**IV** film presents an electrical conductivity value lower than that of the PEDOT:PSS film. It is important to remember that this sample was prepared with the highest concentration of reducing agent. Thus, in order to understand this behavior, we simply mixing PEDOT:PSS and reducing agent ($NaBH_4$) in the same molar ratios and conditions employed for each sample presented in table 1. In general, for a concentration of $NaBH_4$ equal to 1.58×10^{-1} M, we observed the same decrease in the electrical conductivity with a value of σ of 5.32×10^{-4} Scm^{-1}, indicating a possible reduction (de-doping) of PEDOT:PSS for high $NaBH_4$ concentration [22] whereas for the other three samples the electrical conductivity was similar to the pristine

PEDOT:PSS. Therefore, for Ag(PEDOT:PSS)-**IV**, the silver nanoparticles contribution is likely not enough to balance the effect of $NaBH_4$ in the electrical conductivity of PEDOT:PSS. However, for concentrations lower than 5.28 x 10^{-2} M, the effect of $NaBH_4$ on PEDOT:PSS is negligible and thus it could be expected that the same behaviour occurs for the corresponding nanocomposites. As a consequence, the fact that the electrical conductivity increases by decreasing the reducing agent concentration cannot be explained on the basis of a de-doping process.

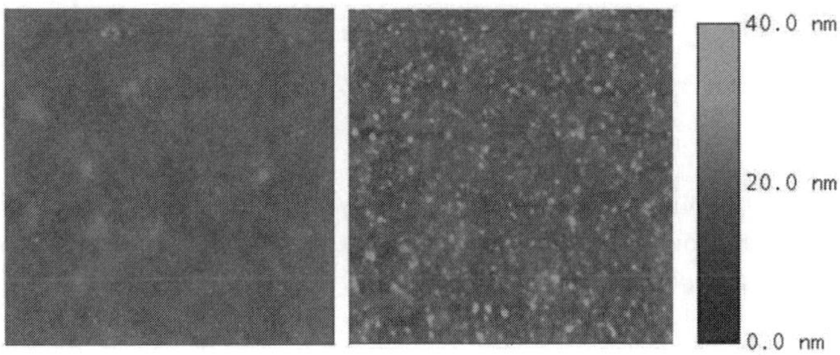

Fig. 3. AFM images of thin films for PEDOT:PSS (left) and Ag(PEDOT:PSS)-**II** (right).

Moreover, for all the reactions carried out for concentration of $NaBH_4$ lower than 5.28 x 10^{-2} M, an enhancement in the electrical conductivity with respect to pristine PEDOT:PSS is obtained, which is another evidence that de-doping can not be evoked in those cases. The electrical behaviour of these nanocomposites must be related with the presence of silver nanoparticles. The trend of the electrical conductivity seems to indicate that bigger particles promote higher electrical conductivity. This result can be actually explained by two hypothetical mechanisms. **1)** The silver nanoparticles formation interferes in the complexation of PEDOT with PSS forcing the poly(thiophene) moieties chains to assume a more extended conformation that is associated with a more conductive quinoid form as suggested by Ouyang et al. [7]. **2)** PSS is an insulating material that interrupts the flux of charge carriers through the conductive PEDOT chains. Several studies have indeed observed that with the removal of the insulating PSS layer, the work function of PEDOT:PSS reduces to a value equal to a highly doped PEDOT [5, 7, 22]. Therefore, the interaction between PSS and silver nanoparticles could overcome this problem, the silver nanoparticles working as electrical connexions between the insulating PSS chains and the conductive PEDOT chains. Even if further studies are necessary to explain the mechanism of conductivity enhancement in PEDOT:PSS by the silver incorporation, this work opens the possibility to manipulate the electrical nature of PEDOT:PSS for specific applications, creating the opportunity for fabricating flexible optoelectronic and electronic plastic devices based on this conjugated polymer [7,8].

Conclusions

We have analyzed four Ag(PEDOT:PSS) composite suspensions prepared with different concentration of $NaBH_4$ as reducing agent. TEM analysis showed that the mean sizes of silver nanoparticles increases when decreasing the concentration of reducing agent. The plasmon absorption peak exhibited dependence with the $NaBH_4$ concentration, tuning toward red when the concentration decreases, which could be related with the increase in the particles diameter as observed by TEM analysis and/or to changes in the dielectric constant of the surrounding medium.

AFM topography showed different morphological properties of thin films of Ag(PEDOT:PSS) composites with respect to PEDOT:PSS film, attributed to the presence of silver nanoparticles in the PEDOT:PSS. It was found an increase in the electrical conductivity of Ag(PEDOT:PSS) films in function of the diameter size of the silver nanoparticles, which could be related to the fact that the stabilization of Ag^+ through PSS chains increases in the charge density when the silver nanoparticles size increases or to conformational change in the polymer induced by the nanoparticles formation.

Acknowledgments

This work has been carried out with the financial support of Mexican Conacyt (project U51504-R) and CNR-CONACyT bilateral program. K.J. Moreno thanks Conacyt for the post doctoral grant as well as CIQA for the project FD0003.

Reference

[1] L.Groenendaal, F. Jonas, D.Freitag, H. Peelartzik, J.R. Reynolds: Adv. Mater. 7 (200) 481.
[2] S. G. Im, K.K. Gleason: Macromol. 40 (2007) 6552
[3] L.A.A. Patterson, S. Ghosh, O. Inganäs: Org. Electron. 3 (2002) 143.
[4] J.Y. Kim, J.H. Jung, D.E. Lee, J. Joo: Synth. Met. 126 (2002) 311.
[5] W. H. Kim, A. J. Mäkinen, N. Nikolov, R. Shashidhar, H. Kim, Z. H. Kafafi: Appl. Phys. Lett. 80 (2002) 3844.
[6] S.K.M. Jönsson, J. Birgerson, X. Crispin, G. Greczynski, W. Osikowicz, A.W.D. van der Gon, W.R. Salaneck, M. Fahlman: Synth. Met. 139 (2003) 1.
[7] J. Ouyang, Q. Xu, C.W. Chu, Y. Yang, G. Li, J. Shinar: Polym. 45 (2004) 8443.
[8] T.K. Sarma, D. Chowdhury, A. Paul, A. Chattopadhyay: Chem. Comm. (2002) 1048.
[9] S.K. Pillalamarri, F.D. Blum, A.T. Tokuhiro, M.F. Bertino: Chem. Mater. 17 (2005) 5941.
[10] T.K. Sarma, D. Chowdhury: J. Phys. Chem. A 108 (2004) 7837.
[11] K. J. Moreno, I. Moggio, E. Arias, I. Llanera, S. E. Moya, R. F. Ziolo, H. Barrientos: J. Nanosci. Nanotechnol. 9 (2009) 3987.
[12] S.S. Kumar, C.S. Kumar, J.Mathiyarasu, K.L.Phani: Langmuir 23, (2007) 3401
[13] D.L.Vav Hyning, C.F. Zukoski: Langmuir 17 (1998) 7034.
[14] A.G. Manoj, Namboothiry, T. Zimmerman, F.M. Coldren: Synth. Met. 157 (2007) 580.
M.A.G. Namboothiry, T. Zimmerman, F.M. Coldren, J. Liu, K. Kim, D.L. Carroll: Synth. Met. 157 (2007), 580.
[15] S. Ghosh, J. Rasmusson, O.Inganäs: Adv. Mater. 10 (1998) 1097.
[16] K.L. Kelly, E. Coronado, Zhao. L.L.Schatz, G.C.: J. Phys. Chem. B 107 (2003) 668.
[17] S.M. Heard, F. Grieser, C.G. Barraclough, V.J. Sanders: J. Colloid Interface Sci.93 (1983) 545.
[18] J.C. Gustafsson, B. Liedberg, O. Inganas, Solid State Ionics 69 (1994) 145.
[19] D.D. Evanoff Jr., G. Chumanov, Chem. Phys. Chem. 6 (2005) 1221.
[20] P.V. Kamat, J. Phys. Chem. B 106 (2002) 7729.
[21] J.J. Mock, D.R. Smith, S. Schultz, Nano. Lett. 3 (2003) 485.
[22] L. Groenendaal, G. Zotti, F.Jonas: Synth. Met. 118 (2001) 105.

Materials Science Forum Vol. 644 (2010) pp 91-95
© (2010) Trans Tech Publications, Switzerland
doi:10.4028/www.scientific.net/MSF.644.91

ELECTRON DIFFRACTION STUDY OF PENTAGONAL CROSS-SECTIONS NANOWIRES

J. Reyes-Gasga[1], J. M. Montejano-Carrizales[2] and M. J. Yacamán[3]

[1] Instituto de Física, Universidad Nacional Autónoma de México, Apartado Postal 20-364, 01000 México, D.F., México.
[2]Instituto de Física, Universidad Autónoma de San Luís Potosí, San Luís Potosí, México.
[3] Department of Physics and Astronomy, University of Texas at San Antonio, San Antonio Texas. 78249-1644
jreyes@fisica.unam.mx

Keywords: Silver nanowires, electron microscopy, electron diffraction patterns.

Abstract. The selected area electron diffraction patterns and structure of pentagonal cross-section nanowires reported for silver are commented and interpreted in is work. We show that they are closely related to a decahedron-base structure.

Introduction

One of the most interesting features of silver nanowires is their structure: they have a pentagonal cross-section [1-5]. Then, it has been proposed that they evolved from a multi-twin decahedral nanoparticle growing in the [110] direction by stabilizing more effectively the {100} facets than the {111} facets. However, up today there is not a complete explanation of all the features presented in their selected area electron diffraction (SAED) patterns, mainly those obtained from their long axis. Note that in these nanowires two different SAED patterns are obtained by tilting 18° from each other. Herein, we present comprehensive electron diffraction analysis on the structure of these penta-twinned nanowires (Dh-NWs).

Pentagonal arrangement in multiple twinned particles (MTP) of transition metals with face-centered cubic (FCC) lattice is quite known [6, 7]. Based on these studies, the basic structure of a decahedral particle was described as the junction of five tetrahedron single crystals with twin-related adjoining faces. By joining 5 tetrahedrons, which are bounded by {111} facets, a gap of ~7.5° is generated. Thus, to fill this gap some form of internal strain is necessary, giving place to dislocations and other structure defects [6-9].

Therefore, the structure of the Dh-nanowires has been described as five triangular prisms joined in such a way that the {100} planes are on their sides and capped by {111} planes, growing along the [110] direction [1, 4-5, 10]. However, up to now it is not clear the actual physical mechanism that allows the growth of such structure and a comprehensive understanding of the structure is required to elucidate the growth mechanism.

Recently, Reyes-Gasga et al. [11] have commented on the interpretation of the contrast observed in the high resolution transmission electron microscope (HRTEM) images of this type of nanowire. In this work, we discuss about their structure through the analysis of SAED patterns from silver nanowires, but all said in this case is also valid for gold and copper Dh-nanowires.

Experimental Procedure

The synthesis for silver nanowires has been reported elsewhere [11]. Silver nanowires were synthesized by the polyol reduction of silver nitrate ($AgNO_3$) in presence of PVP. Transmission electron microscopy was conducted in a HRTEM JEOL 2010F microscope. For digital image processing the Digital Micrograph (GATAN) software was used. To simulate the electron diffraction patterns (the diffractograms), the SimulaTEM software [12, 13] was employed.

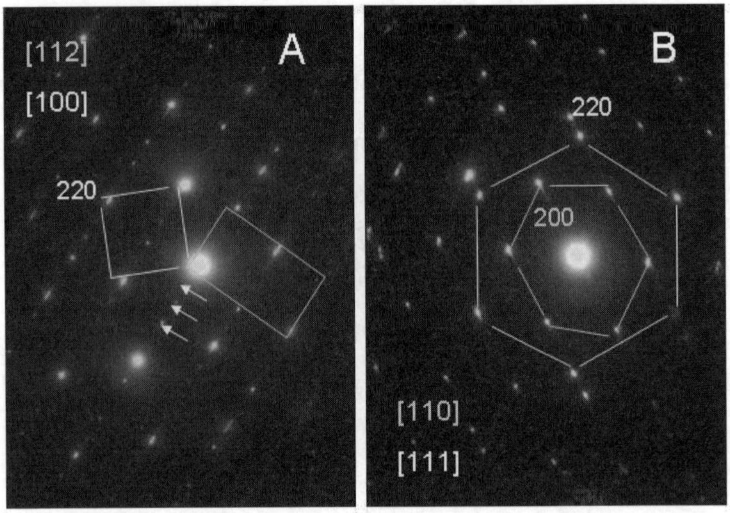

Figure 1. SAED patterns of the Dh-NW. Their indexation indicates that they correspond to the overlapping of [100] and [112] zone axes in (A), and the overlapping of [111] and [110] zone axes in (B). Note the existence of an aperiodic sequence of diffraction spots (indicated by arrows) in (A).

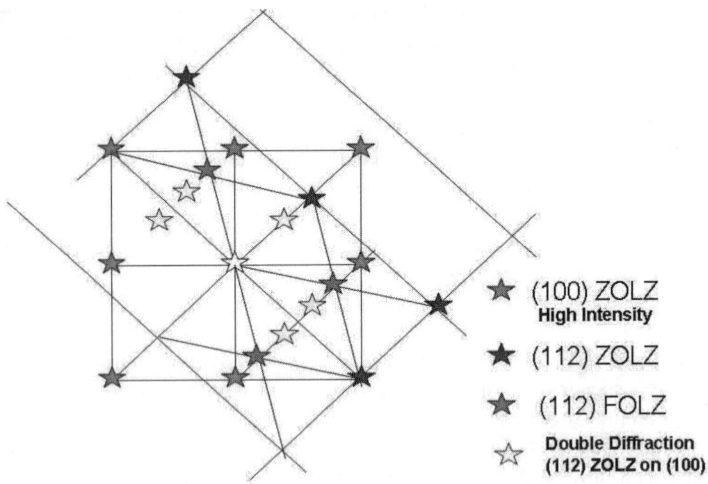

Figure 2. Analysis of the diffracted spots observed in figure 1A. Some of them are ZOLZ spots while others are FOLZ and double diffraction spots.

Results and Discussion

We have already shown that the HRTEM images of the silver nanowires can be interpreted as a Moire pattern contrast based on a penta-twinned decahedron and that the SAED patterns can be also completely generated through the same decahedron basis [11]. In this case, we make the analysis of the origin of their SAED pattern reflections.

SAED patterns along the long axis of individual pentagonal nanowires are presented in Figure 1. Both SAED patterns are easily interpreted as the overlapping of the [100] and [112] zone axes of the silver FCC unit cell in the case of Figure 1a, and the overlapping of the [111] and [110] zone axis for Figure 1b. The presence of an aperiodic sequence of diffraction spots in Figure 1a is of notorious relevance.

The analysis of SAED pattern shown in Figure 1A is presented in figure 2. Because the nanometric size of the nanowire along its cross-section, which produce enlarged reciprocal spots, it is easily deduced that some of the spots belong not only to the Zero Order Laue Zone (ZOLZ) but also to the First Order Laue Zone (FOLZ). Some others are produced by double-diffraction effect.

There are different ways to build an elongated pentagonal dipyramid of the Dh-NW. One consists of five triangular prisms in such a way that they present {100} planes on their sides capped by {111} planes, and growing along the [110] direction (1). We generated this model and obtained its corresponding diffractograms using the SimulaTEM software [12, 13]. The results are presented in Figure 3; note the similarity between these diffractograms and those shown in figure 1.

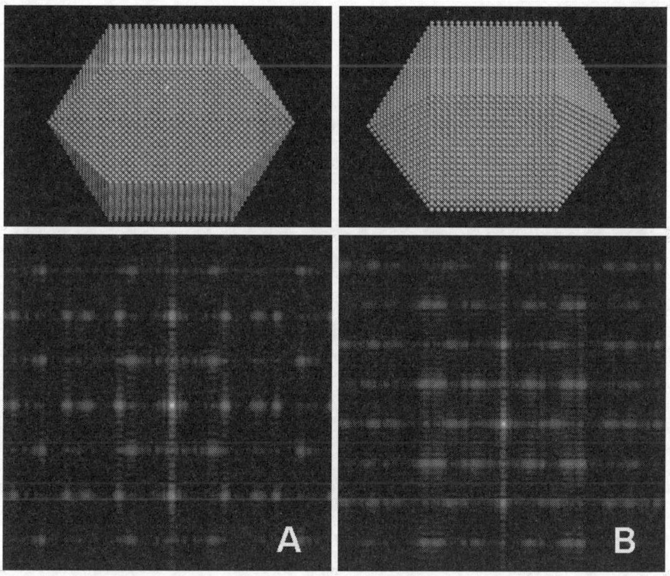

Figure 3. Model of a Dh-NW structure consisting of 5 triangular Ag FCC prisms grew along the pentagonal axis and their corresponding diffractograms. A) Structure along the long axis. B) The same structure as in (A) but rotated by 18° around the five-fold pentagonal axis.

Making the comparison among the diffractograms and experimental diffraction patterns, we noted that although there is presence of an aperiodic sequence of diffracted spots, this sequence is not completely equal to the observed in the experimental patterns. Therefore, the model presented in Figure 3 does not reproduce all the features of the experimental SAED patterns. Therefore, an alternative idea is necessary.

We generated a silver decahedron and simulate the corresponding diffractograms (Figure 4). The comparison between the diffractograms of the 5-twinned silver decahedron and the experimental SAED patterns of the Dh-NW presented in Figure 1 demonstrates that there is an almost perfect match.

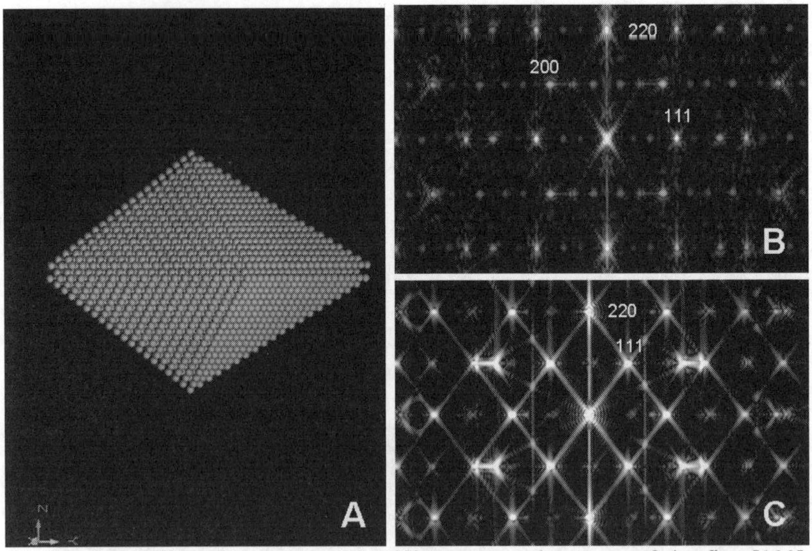

Fig. 3. A) Simulated silver decahedron. B) Diffractogram along one of the five [100] edges. (C) Diffractogram after rotating the direction of (B) by 18° around the five-fold pentagonal axis.

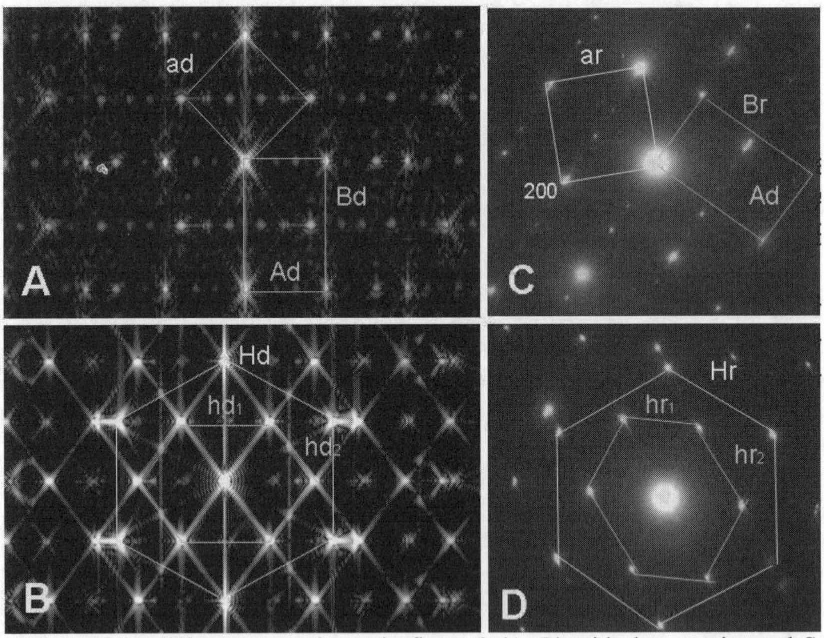

Fig. 4. Comparison of the diffractograms shown in figure 3 (A, B) with the experimental SAED patterns (C, D). The aperiodic sequence observed in the experimental patterns is fully reproduced.

Nevertheless, the comparison of the diffractograms of the decahedron with the experimental SAED patterns (Figure 4) indicates that, making ar/ad = 1.0 and Hr/Hd = 1.0, the ratios Ar/Ad = 0.97 (i.e. 3% of difference), Br/Bd = 0.98 (2%), hr_1/hd_1 = 0.98 (2%), and hr_2/hd_2 = 0.99 (1%). We also noted that the generated aperiodic sequence of diffracted spots is equal to the observed in the experimental patterns. But not only this, *it is similar in positions and in intensity*.

Conclusion

The experimental SAED patterns and the observed diffracted aperiodic sequence is produced by the twinning relationship of the 5 crystals that are forming the decahedral nanostructure and all the images described so far, correspond to a structure based on a 5-twinned decahedron. We propose the structure of the Dh-NWs is produced from a decahedral seed that grow along the vertex (which is parallel to the 5-fold symmetry) by adding pentagonal "caps", in similar way as cones are stacked. In such a model, the Dh-NWs is grown along the [111] direction.

Acknowledgements

The authors gratefully thank to S. Tehuacanero, P. Mexia, M. Monroy and L. Beltran-del-Rio for the technical assistance during this work.

References

[1] Y. Sun, B. Mayers, T. Herricks and Y. Xia: Nanoletters 3 (2003) 955-960.
[2] M. Giersig, I. Pastoriza-Santos, and L. M. Liz-Marzán: J. Mater. Chem. 14 (2004) 607-611.
[3] Q. Zhao, J. Qiu, C. Zhao, L. Hou and C. Zhu: Chem. Letters 34 (2005) 30-31.
[4] H. Hofmeister, S.A. Nepijko, N.N. Ievlev, W. Schulze and G. Ertl: J. Crystal Growth 234 (2002) 773-781.
[5] J. L. Elechiguerra, J. Reyes-Gasga and J. Miguel-Yacaman: J. Mater. Chem. 16 (2006) 3906-3919.
[6] K. Heinemann, M. J. Yacamán, C. Y. Yang and H. Poppa: J. Crystal Growth 47 (1979) 177-186.
[7] Howie and L. D. Marks: Phil. Mag. A 49 (1984) 95-109.
[8] Pauwels, D. Bernaerts, S. Amelinckx, G. Van Tendeloo, J. Joutsensaari, E. I. Kauppinen: J. Crystal Growth 200 (1999) 126-136.
[9] Y. Sun, B. Gates, B. Mayers and Y. Xia: Nanoletters 2 (2002) 165-168.
[10] Y. Gao, P.Jiang, D.F. Lui, H. J. Yuan, X. Q. Yan, Z. P. Zhou, J. X. Wang, L. Song, L. F. Liu, W. Y. Zhou, G. Wang, C. Y. Wang, S. and S. Xie. J. Chem: Phys. Lett. 380 (2003) 146-149.
[11] J. Reyes-Gasga, J. L. Elechiguerra, C. Liu, A. Camacho-Bragado, J. M. Montejano-Carrizales, and M. J. Yacaman: J. Cryst. Growth 286 (2006) 162-172.
[12] Gómez and L.M. Beltrán del Río: SimulaTEM, Revista Latinoamericana de Metalurgia y Materiales. 21 (2001) 46-50.
[13] A. Gómez-Rodríguez, L.M. Beltrán-del-Río y R. Herrera-Becerra, SimulaTEM: Ultramicroscopy, Accepted 2009

Materials Science Forum Vol. 644 (2010) pp 97-100

doi:10.4028/www.scientific.net/MSF.644.97

Study on the Microstructure and Electrical Properties of $Pb(Zr_{0.53}\ Ti_{0.47})O_3$ Thin-films

L. A. Baldenegro-Perez*[,a], W. Debray Mcchtaly[b], E. Fuentes-Fernandez[c], M.A. Quevedo-Lopez[d], H.N. Alshareef[e], Pradeep Shah[f] and B.A. Gnade[g]

Department of Materials Science and Engineering, The University of Texas at Dallas, 800 West Campbell Rd., 75080,Richardson, TX, USA

[a]leobaldenegro@yahoo.com, [b]wardia.debray@gmail.com, [c]eemf081000@utdallas.edu, [d]mquevedo@utdallas.edu, [e]husam.alshareef@kaust.edu.sa, [f]pshah972@yahoo.com, [g]beg031000@utdallas.edu

Keywords: PZT, thin-films, sol-gel, piezoelectric, electrical characterization.

Abstract. In the present study a complete analysis of the morphological and electrical properties of PZT layers with composition 53Zr-47Ti is presented. Three different samples composed of 3, 6, and 9 PZT layers were analyzed on a substrate consisting of ZrO2-SiO2-Si structures. The PZT and ZrO2 layers were deposited via Sol-Gel, whereas the SiO2 layer, on every sample, was deposited via PECVD. SEM results showed morphology of very small granules on the 3 layered thin-film samples (12 nm), on the 6 layered thin-film samples a mixture of small and large size (100-300 nm) granule formation was observed, with the 9 layered thin-film samples exhibiting very large granule sizes (bigger than 300 nm). XRD results showed that increasing the number of deposited layers caused an incremental increase on the detected peak intensities, aided in the promotion of the perovskite phase, and diminished the presence of the pyrochlore phase. It was also observed, during electrical measurements, that increasing the number deposited layers directly increased the overall capacitance of the thin-film structure. This effect was attributed primarily to the large amount of perovskite and large size of grains presented on thick samples.

Introduction

Lead Zirconate Titanate (PZT) can provide actuation and sensing properties, this material also presents higher piezoelectric constants because the dipoles are easily aligned by relatively low electric fields. For these reasons, PZT have been vastly investigated for pyroelectric and piezoelectric sensors, non-volatile memory, high frequency tunable devices, micromechanical systems (MEMS) such as micro accelerometers, micro motors and infrared sensors among other applications [1].

Ferroelectric thin films can be deposited by traditional techniques such as sol-gel, magnetron sputtering and dip-coating, or non-traditional techniques such as metal organic chemical vapor deposition (MOCVD), electron cyclotron resonance plasma enhanced CVD (ECR PECVD) method and pulsed laser deposition (PLD) [2,3].

In addition, PZT thin films need to have a proper buffer layer to prevent or minimize interface reactions between Pb and Si or SiO_2 substrates that produce deterioration in ferroelectric properties. Also, it was demonstrated that a buffered layer can be effectively utilized to engineer the final properties of the film stack [4], because allows film orientation and microstructures, to control growth rate and film nucleation. Some buffered layer recipes have been reported in the literature such as Zirconia (ZrO_2), TiO_2, Lead Titanate ($PbTiO_3$ or PT) and $LaNiO_3$ [4].

In the present study, a sol-gel ZrO_2 process was used as a buffered layer for chemical durability, acceptable mechanical strength, alkali resistance, low cost, ease of production and deposition by spin-coating on SiO_2/Si substrates [5]. The effect of ZrO_2 on the PZT morphology and electrical properties were investigated as a function of PZT thickness.

Experimental Procedure

A sol-gel solution (0,5 mol/l) was used to deposit a ZrO_2 thin-film onto SiO_2/Si substrate. Zirconium (IV) butoxide $Zr[O(CH_3)_3CH_3]_4$ was used as a precursor. A sol-gel solution was preparared by dissolving the Zirconium (IV) butoxide in isopropanol and acetic acid and finally water was added to complete the hydrolysis, as were reported by Lima-Neto et al. [6]. The solution was then spin coated for 30 sec at 3000 rpm (~40nm). A hot plate at 300°C was used to pyrolize the organic compounds and solvents and then samples were crystallized at 800°C for 30 minutes on air.

A PZT and Lead Titanate (PT) sol-gel 0.4 mol/l were prepared. The PZT sol-gel process used was previously reported by Assink and Schavartz [7]. The Zr and Ti ratio was 0.53/0.47 respectively (20% excess of Pb). It has been reported that this ratio presents excellent electrical properties at this composition in bulk PZT samples due the presence of nanodomain structures and tetragonal microdomains [8]. Also, a PT solution was made and used as a nucleation layer in order to improve the morphology of PZT grains. All samples were crystallized at 650°C for 30 min. Three different samples were created, the first one with 2PZT-1PT, the second one with 3PZT-2PZT-1PT and the third one with 3PZT-3PZT-2PZT-1PT configuration, labeled as 3PZT, 6PZT and 9 PZT samples respectively. The thickness of the 3PZT sample was approximately from 90 nm.

SiO_2 deposition was made by PECVD on a Silicon substrate (Unaxis 790). The morphology of ZrO2, PT and PZT films were analyzed using a SEM (Zeiss Supra-40). The crystalline phases were determined using XRD (Rigaku, CuKα radiation). Capacitance measurements (Capacitance vs Voltage) in the range of 10KHz-100 KHz and 1MHz were obtained using a Cascade Microtech Probe-Station (SUMMIT, 11861B) and an impedance analyzer (Hewlett Packard-Agilent HP484-CFG010). A surface profilometer (Veeco Detak VIII) was used to analyze the thicknesses.

After PZT depositions, a lift off process was used to generate interdigital electrodes (IDE) with fingers and gaps between fingers both of 2 μm in dimension (25 electrodes). A metal evaporator (Temescal 1800 e-beam evaporator) deposited Au (100nm)/Cr(10nm) electrodes.

Results and discussion

An SEM picture and a XRD pattern of a ZrO_2 thin film deposited on a SiO_2/Si substrate is shown in Figure 1. The ZrO_2 morphology showed grains from 6 to 10 nm of average size. The cubic and tetragonal peaks also appear in the XRD spectrum, but monoclinic phase predominates at annealing temperature of 800°C. These behaviors were reported by Chatterje et al. [9], not for thin-films, but for particles prepared by precipitation and calcination at the same temperature. They also found a grain size of 13 nm by using Scherrer equation in conjunction with measured XRD spectra.

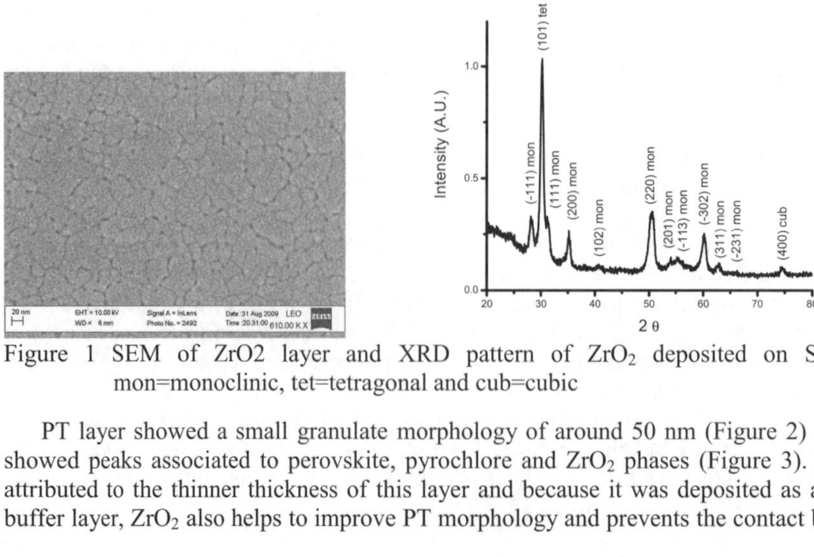

Figure 1 SEM of ZrO2 layer and XRD pattern of ZrO_2 deposited on SiO_2/Si substrate, mon=monoclinic, tet=tetragonal and cub=cubic

PT layer showed a small granulate morphology of around 50 nm (Figure 2) and XRD pattern showed peaks associated to perovskite, pyrochlore and ZrO_2 phases (Figure 3). ZrO_2 peaks were attributed to the thinner thickness of this layer and because it was deposited as a first layer. As a buffer layer, ZrO_2 also helps to improve PT morphology and prevents the contact between SiO_2 and the ferroelectric material. The PZT sample (3 PZT) showed a very small granulate structure (~12 nm) and was associated with high and low levels of perovskite and pyrochlore phases, respectively. Also, a cross section of the 3 PZT sample reveals a ~90 nm thickness for this stack (Figure 3). On the other hand, a mixture of very small and large grains (from 100 to 300 nm) was observed on the 6 PZT sample. These large grains were associated to the perovskite phase only. This behavior is well in agreement with the decrement of the pyrochlore phase observed in the XRD spectra as the thickness of the sample was increased. The 9 PZT sample showed large grains (bigger than 300 nm) and apparently were produced from collapsing and assemblage of small and medium grains.

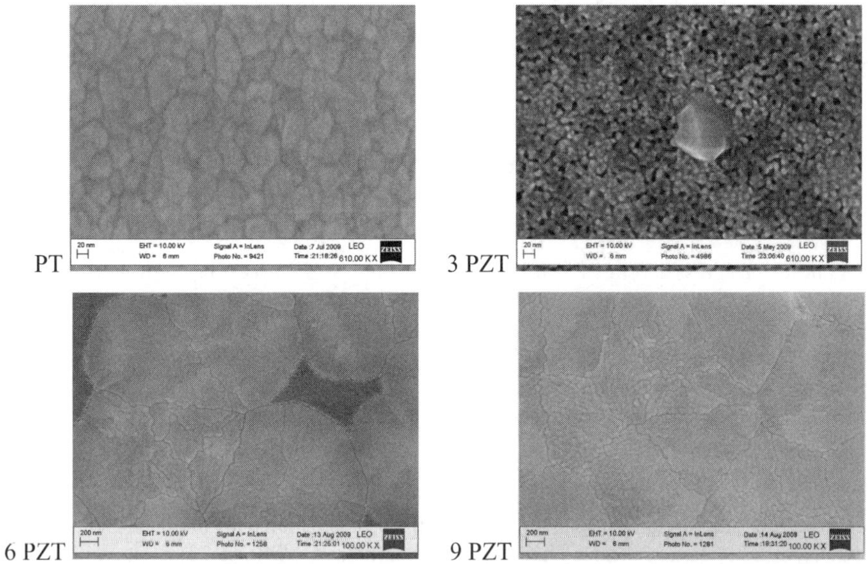

Figure 2 SEM of PT layer and 3, 6 and 9 PZT samples deposited on $ZrO_2/SiO_2/Si$ substrate

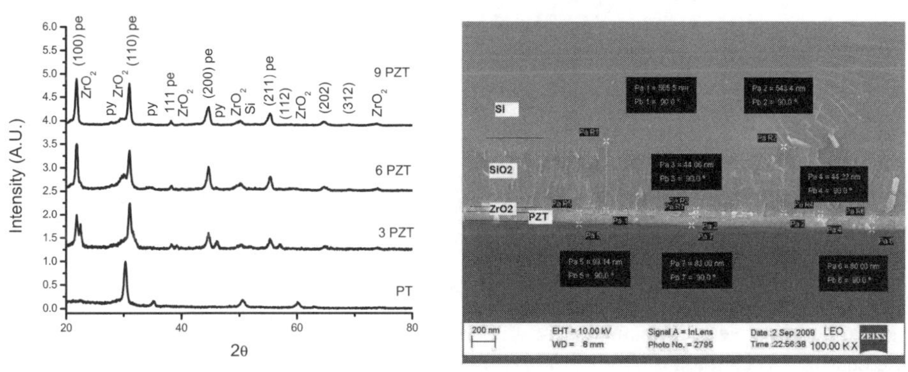

Figure 3 XRD patterns of PT, 3, 6 and 9 PZT samples deposited on $ZrO_2/SiO_2/Si$ substrate and SEM cross section of 3PZT sample, pe=perovskite, py=pyrochlore, Si=Silicon [3]

In order to analyze the electrical properties of these three samples, a study of Capacitance vs Voltage was presented in Figure 4. For every sample, three different frequencies were used (10, 100 KHz and 1 MHz). It can be observed in all cases, from 3 PZT to 9 PZT samples that the capacitance decreased when frequency was increased. This behavior can be associated to a delayed dipole-field alignment at high frequency. On the other hand, an increment of the PZT thickness produced an increment of the capacitance and this effect can be attributed to: a) a perovskite phase dominates on thicker samples and also b) an increment on grain size improves electrical properties [10]. A simultaneously increase of capacitance and thickness has also been reported on other ferroelectric material [(Ba, Sr) TiO_3][11]. These authors attributed this increment to the change in film strain, the contraction in film lattice and the reduction of interfacial dead layers between films and electrodes.

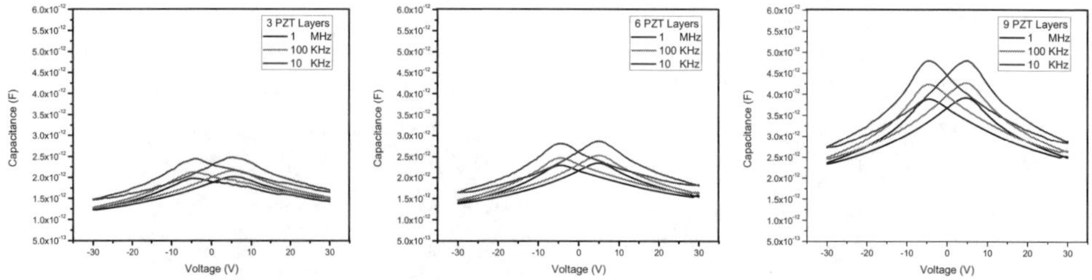

Figure 4 Measured Capacitance vs Voltage for 3 PZT, 6 PZT and 9 PZT samples

Conclusions

Morphological and electrical properties on different PZT samples (53 Zr/47 Ti) deposited on ZrO_2 buffered layer were investigated. Samples of 3 PZT, 6 PZT and 9 PZT showed small grains (~12 nm), large grains (100 - 300nm) and very large grains (> 300nm), respectively. These large grains were associated to perovskite phase from XRD analysis. Thicker PZT samples showed and increment of the measured capacitance and this effect were attributed to a predominant perovskite phase and large size of grains on thick samples.

Acknowledgments

L. A. Baldenegro-Perez acknowledge financial support from the National Council on Science and Technology (CONACYT, Mexico) under Posdoctoral Scholarship (75930, 92707). The authors are grateful to Texas Piezoelectric, Inc. and Gabriel Padron-Wells from the University of Texas at Dallas for their help in this study.

References

[1] D.L. Polla, L.F. Francis: MRS Bull., Vol. 21 (1996), p. 59

[2] Eunki Hong, S.V. Krishnaswamy, C.B. Freidhoff and S. Trolier-McKinstry: Sensors and Actuators A, Vol. 119 (2005), p. 520

[3] J. Wang, ZG. Wu, X.M. Yuan, S.R. Jiang, P.X. Yan: Mat. Chem. Phys., Vol. 88, (2004), p. 77

[4] Taeyum Kim, A. I. Kingon, Jon-Paul Maria: J. Am. Ceram. Soc., Vol. 89-11, (2006), p. 3426

[5] M. Atik and M.A. Aegerter: J. Non-Cryst. Solids, Vol. 147, (1992), p. 813- 819

[6] Pedro de Lima Neto, Mahamed Atik, Luis A. Avaca and Michael A. Aesorker: Journal of Sol-Gel Science and Technology, Vol. 1 (1994), p. 177

[7] R. A. Assink and R. W. Schwartz: Chem. Mater., Vol. 5 (1993), p. 511

[8] K.A. Schonau, M. K, H. K., M.J. Hoffmann and H. F.: Phys. Rev. B, Vol. 76 (2007), p. 144112

[9] A. Chatterjee, S.Pradhan, A.Datta, M.De, D.Chakraverty: J. Mater. Res., Vol. 9 (1994), p. 263

[10] M.H. Frey and D.A. Payne: Phys. Rev. B: Condens Matter., Vol. 54 (1996), p. 3158

[11] J.Y. Ha, J.W. Choi, C.Y. Kang, S.F. Karmanenko, D.J. Choi, S.J. Yoon and H.J. Kim: J Electroceram, Vol 17 (2006), pag 141-144

Materials Science Forum Vol. 644 (2010) pp 101-104

doi:10.4028/www.scientific.net/MSF.644.101

Formation of Si nanocrystals in thin SiO_2 films for memory device applications

Mario Curiel[1,a], Ivan Petrov[2,b], Nicola Nedev[1,c], Diana Nesheva[3,d], Mauro Sardela[2,e], Yuya Murata[2,f], Benjamin Valdez[1,g], Emil Manolov[3,h] and Irina Bineva[3,i]

[1]Institute of Engineering, Autonomous University of Baja California, Benito Juarez Blvd. esc. Calle de la Normal, s/n, C. P. 21280 Mexicali, B. C., Mexico

[2]Center for Microanalysis of Materials, Frederick Seitz Materials Research Laboratory, University of Illinois, Urbana, Illinois 61801, USA

[3]Institute of Solid State Physics, Bulgarian Academy of Sciences, 72 Tzarigradsko Chaussee Blvd, 1784 Sofia, Bulgaria

[a]marioc@iing.mxl.uabc.mx, [b]petrov@illinois.edu, [c]nicola@iing.mxl.uabc.mx, [d]nesheva@issp.bas.bg, [e]sardela@illinois.edu, [f]muratacarbon@gmail.com, [g]benval@iing.mxl.uabc.mx, [h]emman@issp.bas.bg, [i]irka@issp.bas.bg

Keywords: SiO_x, Si nanocrystals, TEM, AFM.

Abstract. X-ray Diffraction and Reflectivity, Transmission Electron Microscopy and Atomic Force Microscopy were applied to study the effect of thermal annealing on the properties of thin SiO_x films (~ 15 nm) prepared by thermal evaporation of SiO in vacuum. It has been shown that furnace annealing at 1000 °C causes phase separation and formation of uniformly distributed Si nanocrystals into a SiO_2 matrix. Clockwise hysteresis has been observed in the C-V curves measured and explained by assuming charging and discharging of the NCs with carriers, which tunnel from the Si substrate.

Introduction

Silicon dioxide films containing Si nanocrystals (Si NCs) have attracted much attention in the recent years because of the possibility to use them in Si-based light emitters and in non-volatile memory devices [1,2]. These potential applications require good knowledge of the structural, compositional and electrical properties of the SiO_x films deposited by different techniques. In this work we present results about the properties of thermally evaporated SiO_x films with initial compositions of x = 1.1 and 1.3 obtained by Transmission Electron Microscopy (TEM), Atomic Force Microscopy (AFM) and X-ray Diffraction (XRD) and Reflectivity (XRR). Metal-Insulator-Semiconductor (MIS) structures containing 'fresh' or annealed SiO_x film in the gate stack were characterized electrically.

Experimental details

SiO_x films (~15 nm thick) with initial composition x = 1.1 and 1.3 were fabricated by thermal evaporation of SiO [3,4] on top of an n- or p-type crystalline silicon (c-Si) at a vacuum of 1×10^{-3} Pa. Silicon nanocrystals were grown by annealing of the films at 1000 °C in N_2 atmosphere for 60 min. Structural characterization using XRD, XRR and AFM was carried out on the annealed films as well as on control ones, annealed at 250 °C for 30 min only for stabilization. Lattice-resolution transmission electron microscopy was carried out at 200 kV with a JEOL 2100 with a point-to-point resolution of 0.22 nm in the TEM mode.

Test Metal-Insulator-Semiconductor structures containing Si NCs in the gate dielectric were fabricated by evaporation of a SiO_x (x = 1.1 or 1.3) layer with a thickness of ~15 nm on top of a n-type c-Si (4-6 Ω cm), followed by radio frequency sputtering of a silicon dioxide film with thickness between 30 and 40 nm. Half of the structures were annealed at 1000 °C for 60 min.

Aluminum was evaporated through a mask and capacitors with areas of 2×10^{-3} cm^{-2} were formed. The MIS structures were characterized electrically by Capacitance–Voltage (C-V) measurements.

Results and discussion

X-ray diffraction showed that the control samples as well as the ones annealed at 1000 °C are amorphous. The XRR technique was used to measure the thicknesses of the SiO_x layers before and after annealing and values in the range 15 - 17 nm were obtained. The annealing leads to a small decrease (≤10%) of the initial film thickness for all samples, which is a result of SiO_x film densification. The values for the thickness derived from XRR are in good agreement with those determined during the deposition. The amplitude of the interference fringes measured by XRR is smaller in the annealed samples than in the control ones, which implies increase of the c-Si wafer/SiO_x interface roughness after annealing.

Figures 1 (a) and (b) show plan-view and cross-sectional (XTEM) micrographs of a control SiO_x film with x = 1.3. In the XTEM image (Fig. 1 (b)) nearly atomically flat interface (~ 2 monolayers) between the crystalline Si wafer and the dielectric layer is seen. The value of 17 nm for the SiO_x film thickness determined from XTEM is close to the one obtained by XRR. Figures 2 (a) and (b) show plan-view and cross-sectional TEM images of annealed films with the same composition.

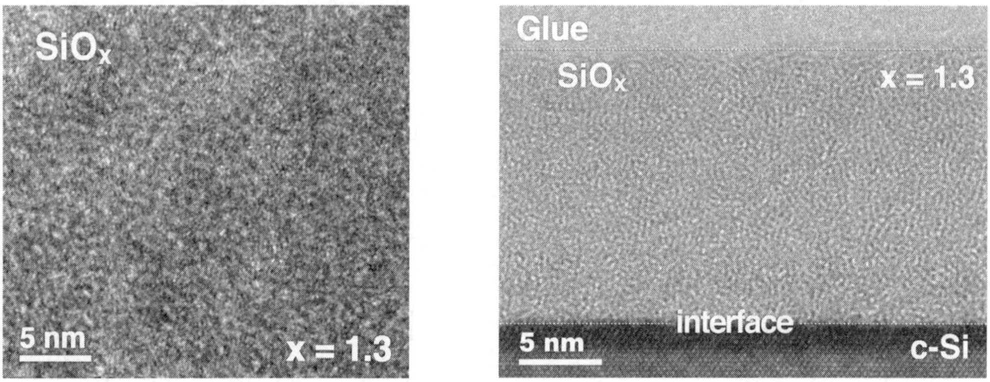

(a) (b)

Fig. 1. Plan-view (a) and cross-sectional (b) TEM images of a control SiO_x film with x = 1.3.

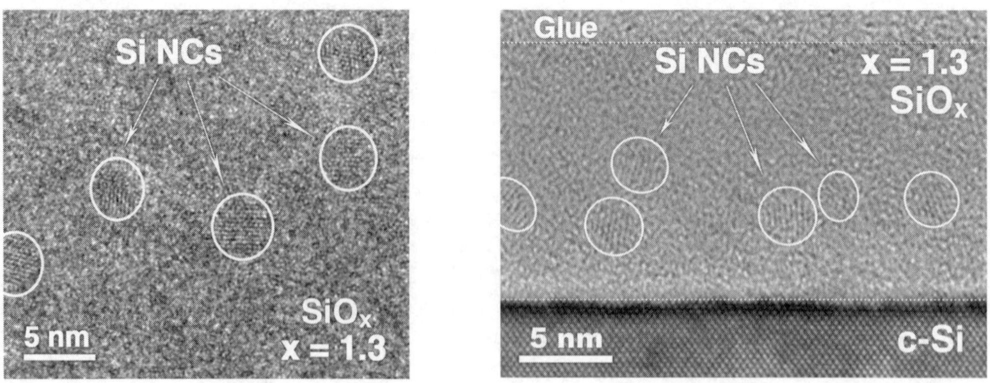

(a) (b)

Fig. 2. Plan-view (a) and XTEM (b) images of SiO_x (x = 1.3) films annealed at 1000 °C for 60 min.

Formation of uniformly distributed NCs with diameter of about 4-5 nm and average distance between them of ~ 4-5 nm (Fig. 2(a)) is observed. The distance between the NCs and the c-Si substrate is ~3 nm (Fig. 2 (b)). XTEM supports the XRR results about the decrease of the thickness of annealed film (~15 nm) and the smoother interface in the control sample. High temperature annealing increases the roughness of the c-Si/SiO_x interface to 3-4 monolayers (Fig.2 (b)).

The higher content of excess Si in the SiO_x films with initial composition of x = 1.1 leads to an increase of the nanocrystal size to 5 – 6 nm (Figs. 3 (a), (b)). In figure 3 (b) the distance between the Si wafer and the NCs is ~ 4 nm.

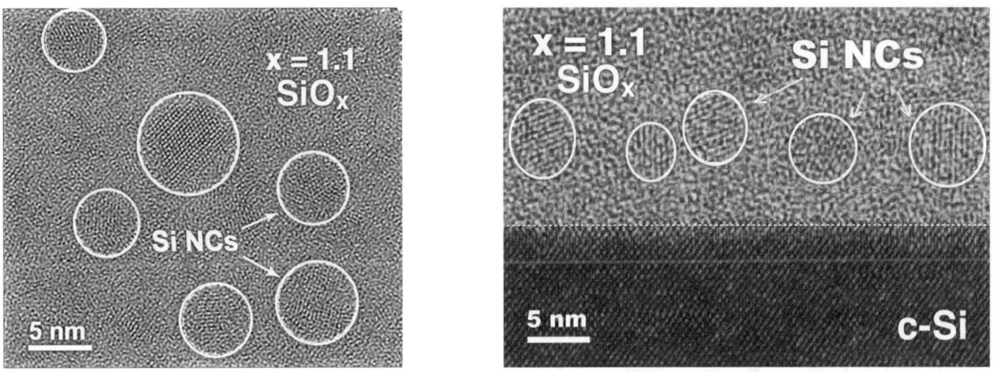

(a) (b)

Fig. 3. Plan-view (a) and XTEM (b) images of SiO_x (x=1.1) films annealed at 1000 °C for 60 min.

The AFM images in figures 4 show the morphology of a SiO_x layer with x = 1.3 before (a) and after annealing (b). The AFM measurements revealed that the annealing leads to a decrease of the surface roughness for both initial compositions, a result, which is important for incorporation of the films in non-volatile memories. The value of the roughness (RMS) obtained by AFM is 0.255 nm for the control (Fig. 4 (a)) and 0.164 nm for the annealed sample (Fig. 4 (b)).

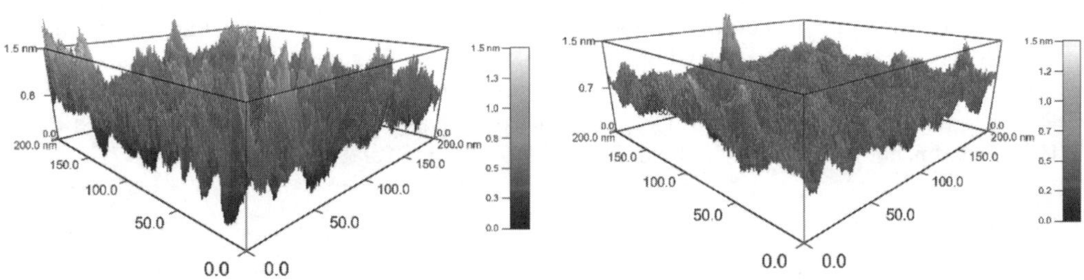

(a) (b)

Fig. 4. AFM images of a control (a) and an annealed (b) SiO_x layer with x = 1.3.

Figures 5 (a), (b) show the C-V hysteresis curves of test MIS structures with annealed SiO_x films having x = 1.1 and 1.3, respectively. The curves were measured in two voltage ranges, ±15 (curve 2) and ±18 V (curve 3, Fig. 5 (a)), and ±6 (curve 2) and ±8 V (curve 3, Fig. 5 (b)). Here ± scanning means that the gate bias is swept in both directions. Similar measurements were carried out on control MIS structures and no such parallel shifts were observed. The clockwise hysteresis seen can be explained by assuming charging and discharging of the NCs with carriers, which tunnel from the

Si substrate. The slope of the curves in Fig. 5 (b) is higher, which implies [5] better c-Si/dielectric interface for the film with x = 1.3. The charging of the x = 1.3 structure can be done with lower gate voltage but its retention characteristics are worse than those of the x = 1.1 structure. For 10^5 s the value of the charge trapped in the NCs decreases to 20 %, 50 % for the x = 1.3, 1.1 structure, respectively.

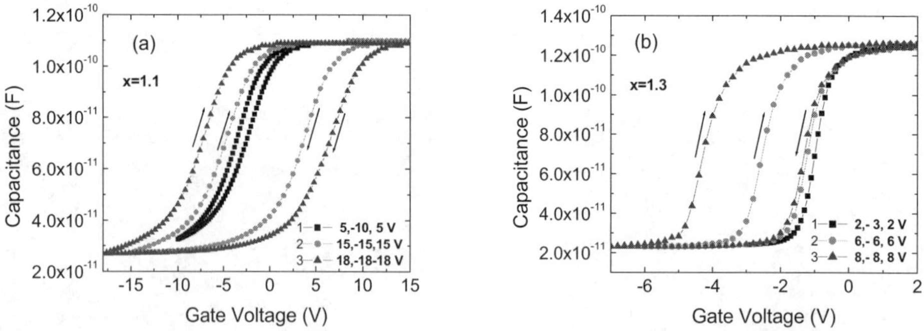

Fig. 5. C-V dependences of annealed structures containing SiO_x film with x = 1.1 (a) and x = 1.3 (b) measured at 1 MHz. Curve 1 is the initial one, while curves 2 and 3 are scanned in the ranges ±15, ±18 V (a) and ±6, ±8 V (b).

Conclusion

It has been shown that furnace annealing of SiO_X (x = 1.1 and 1.3) films at 1000 °C causes phase separation, film densification and small change of the c-Si/SiO_X interface. Uniformly distributed Si nanocrystals embedded into a SiO_2 matrix and separated by an oxide layer from the c-Si wafer (~ 3-4 nm thick) has been proven by TEM measurements. Higher Si content leads to an increase of the NC size. Clockwise hysteresis has been observed in the C-V curves measured and explained by assuming charging and discharging of the NCs with carriers, which tunnel from the Si substrate.

Acknowledgements

The authors gratefully acknowledge the financial support of Consejo Nacional de Ciencia y Tecnologia CONACyT, Mexico and the Bulgarian Ministry of Education and Science under grants NT-4-1 and F1504. The measurements were carried out in part in the Frederick Seitz Materials Research Laboratory Central Facilities, University of Illinois, which are partially supported by the U.S. Department of Energy under grants DE-FG02-07ER46453 and DE-FG02-07ER46471.

References

[1] S. Ossicini, L. Pavesi, F. Priolo, in: Light Emitting Silicon for Microphotonics, (Springer tracts in modern physics), Vol. 194, Springer-Verlag, Berlin (2003)

[2] S. Tiwari, F. Rana, H. Hanafi, A. Hartstein, E. Crabbé and K. Chan: Appl. Phys. Lett. Vol. 68, (1996), p. 1377

[3] D. Nesheva, C. Raptis, A. Perakis, I. Bineva, Z. Aneva, Z. Levi, S. Alexandrova, H. Hofmeister: J. Appl. Phys. Vol. 92 (2002), p. 4678

[4] D. Nesheva, I. Bineva, Z. Levi, Z. Aneva, Ts. Merdzhanova, J. C. Pivin, Vacuum Vol. 68 (2003), p. 1

[5] S.M. Sze, in: Physics of Semiconductor Devices, chapter 7, 2nd edition, Wiley, NY (1981)

Materials Science Forum Vol. 644 (2010) pp 105-108

doi:10.4028/www.scientific.net/MSF.644.105

Fe_2O_3 Thin Films Prepared by Ultrasonic Spray Pyrolysis

M.A. García-Lobato, A. Hernández-V, H.M. Hdz-García,

Arturo I. Martínez[a], and M. I. Pech-Canul

Centro de Investigación y de Estudios Avanzados del Instituto Politécnico Nacional (Cinvestav-Saltillo), Carr. Saltillo-Monterrey Km. 13 C.P. 25900, Ramos Arizpe, México.

[a]mtz.art@gmail.com

Abstract. Fe_2O_3 thin films were obtained on glass substrates by the ultrasonic spray pyrolysis technique; additionally, a thermodynamic analysis on the Fe_2O_3 formation from different precursor salts was performed. In this work, the influence of the deposition temperature is related with the structural, optical, and morphological properties of the iron oxide films. The present results reveal that amorphous thin films are obtained at temperatures lower than 350° C, while polycrystalline films with the maghemite structure are formed at higher temperatures.

Introduction

During various decades, iron oxides have been widely studied; more recently, these oxides have been developed in the form of nanopowders and thin films. This is due to the wide variety of applications in various areas of science and technology. These applications include drug delivery vehicles [1], solar filters [2], recording media [3], spin valves [4], electrochromic devices [5], lithium-ion batteries [6], photo-electrochemical system for hydrogen generation [7], among others. There is a great variety of iron oxides, which include those with iron ions in the form of Fe (II) and others with Fe (III); also, some include a mixture of the two valences. Polymorphs with a general formula of Fe_2O_3 are thermodynamically more stable, this allows their use at different conditions. The Fe_2O_3 phases include hematite (α-Fe_2O_3), maghemite (γ-Fe_2O_3), ε-Fe_2O_3, and δ-Fe_2O_3; the α and β phases exhibit the rhombohedral structure, the γ a spinel-like structure, and the ε an orthorhombic structure.

In order to bring out the necessary information for future researchers in the field of iron oxides, the evaluation of appropriate parameters for the synthesis of iron oxide thin films with the highest optical quality is carried out. For the design of solar filters, the aim of this work is to study the influence of different deposition parameters on the physical properties of the Fe_2O_3 thin films prepared by the ultrasonic spray pyrolysis technique.

Thermodynamic Analysis of the Fe_2O_3 formation

In the spray pyrolysis technique, the precursor salts have to fulfill some basic criteria; such as high solubility in a desired solvent, and low decomposition temperature. The method is based in the pyrolytic decomposition of precursors; the following reaction describes the formation of a metal oxide starting from its corresponding salt:

$$MA_{x(s)} + H_2O_{(l)} \rightarrow MO_{y(s)} + xHA_{(g)} \qquad (1)$$

Where M is a metal, and A is an anion. In order to select an adequate precursor salt, a thermodynamic analysis can be conducted using a software package such as HSC 6.12 by Autokumpu. The probability of Fe_2O_3 formation starting from different salts could be expressed in terms of the reaction constant (K); for the iron oxide formation K must be positive. At equilibrium, when K is much larger than 1, the concentration of products is far greater than the reactants. Figure 1 shows the log (K) as a function of temperature, from 0 to 800 °C. In figure 1, it can be observed that iron sulphates (ferrous and ferric) can not form iron oxides at temperatures lower than 430 °C. Also, some iron halides were evaluated; both the fluoride and bromide salts show that the formation of Fe_2O_3 can be feasible at temperatures higher than 230 °C with low *K*. In contrast, it can be observed that both $FeCl_2$ and $FeCl_3$ salts show a lower decomposition temperature than the other precursors; in fact, the ferrous chloride forms the oxide at temperatures as low as 0 °C. Additionally, the iron (III) nitrate can form the oxide at temperatures higher than 180 °C, it can be observed that the log (*K*) growth faster than other precursors.

These thermodynamic results suggest which precursors are not adequate in the preparation of iron oxide films. For example, the $FeCl_2$ can not be used, it is due to that it forms the iron oxide at very low temperatures and it can be unstable in aqueous solution. Also, the iron sulphates are not suitable because of the iron oxide can only be deposited at very high temperatures; it avoids the use of glass substrates. On the other hand, the use of bromides and fluorides do not offer any thermodynamic advantage vs. the iron (III) chloride, and they are more expensive and toxic; consequently, these precursors will not be used experimentally in this work.

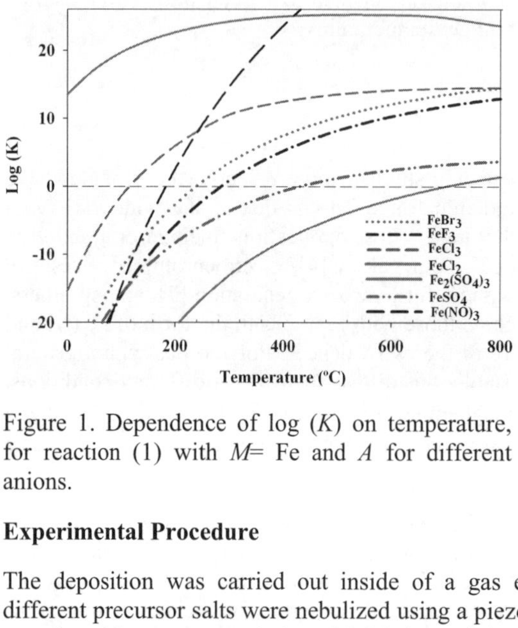

Figure 1. Dependence of log (*K*) on temperature, for reaction (1) with *M*= Fe and *A* for different anions.

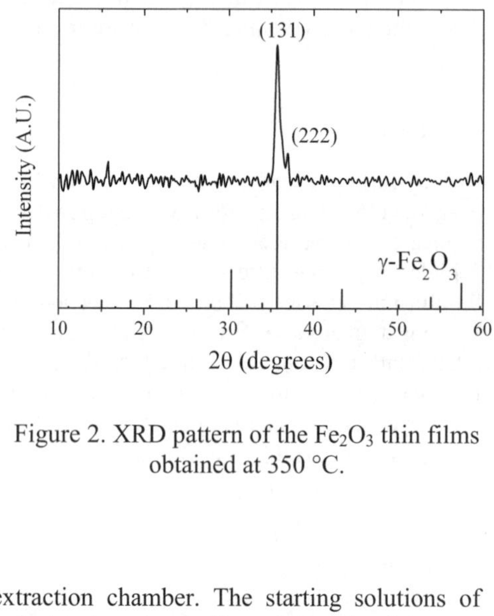

Figure 2. XRD pattern of the Fe_2O_3 thin films obtained at 350 °C.

Experimental Procedure

The deposition was carried out inside of a gas extraction chamber. The starting solutions of different precursor salts were nebulized using a piezoelectric transducer. The solution droplets were transferred by compressed air at 5.9 L/min, and the deposition temperatures range from 200 to 400 °C. At these conditions, the consumption of the solution is invariable at a flux of 0.36 ml/min. The substrate temperature was regulated using a temperature controller (Omron E5CN); it was kept constant during the deposition time (120 s). In order to decrease the residual stress of the film/substrate system, the coated substrates were cooled slowly until reaching ambient temperature.

In agreement with the previous section, 0.5 M aqueous solutions of the $FeCl_3$ and $Fe(NO_3)_3$ salts were used in the preparation the iron oxide films. However, aqueous solutions of $FeSO_4$ were also sprayed; according to the thermodynamic analysis, no films were obtained at the deposition temperatures (200-400 °C), it is due to the negative *K* values (see figure 1). Additionally, when solutions of iron nitrate are sprayed, both very low deposition rates and yields were obtained. It may be due to the very high *K* values, which promotes the formation of the iron oxide before the precursor droplets reach the substrate surface. The best deposition rates and yields were obtained for aqueous solutions of iron (III) chloride; therefore, only the deposited films starting from this precursor are analyzed in this work.

The morphology of the films were studied by scanning electron microscopy (SEM) using a Philips XL30 ESEM microscope. The crystalline structure was determined by X-ray diffraction (XRD) in a Philips X'Pert diffractometer using both the Brag-Brentano, and the grazing incidence geometries. The optical analysis was performed using a spectrometer (Hach DR5000) in the range UV-VIS-NIR from 190 to 1100 nm.

Results and Discussion

Structure and morphology

The Fe_2O_3 thin films show an excellent adherence to the glass substrates. When using the Brag-Brentano geometry for the XRD tests, all the films show an amorphous structure. However, as shown in figure 2 for films deposited at a temperature of 350 °C, a polycrystalline structure is exhibited by the grazing incidence geometry. The XRD pattern can be indexed with the maghemite structure, PDF file 391346. A high intensity (131) peak can be observed; also, the (222) peak can be indexed to the maghemite structure as it is shown in the figure. As we expected, the deposition temperature is related with an increase of the crystallinity of the films. It is known that higher temperatures promote a high mobility of ad-atoms; it allows a better atomic arrangement, which promotes the formation of a crystalline structure; in our particular case, the maghemite structure.

Figure 3 shows SEM micrographs of iron oxide films deposited at 350 and 450 °C; both films show different surface morphology. As it is evidenced in the figure, films deposited at 450 °C are rougher than films deposited at 350°C. This phenomenon is due to the fact that droplets evaporate before they reach the substrate surface, promoting the precipitation of the precursors. Therefore, the kinetic energy of ad-atoms decreases and forms rough surfaces. On the other hand, at the optimum substrate temperature, the pyrolytic reaction takes place at the gas-substrate surface interface; it promotes the synthesis of smoother films similar to those obtained by a - CVD process. For optimal deposition temperatures, the ultrasonic technique allows to produce homogeneous and smoother films that those obtained by a pneumatic spraying [8, 9].

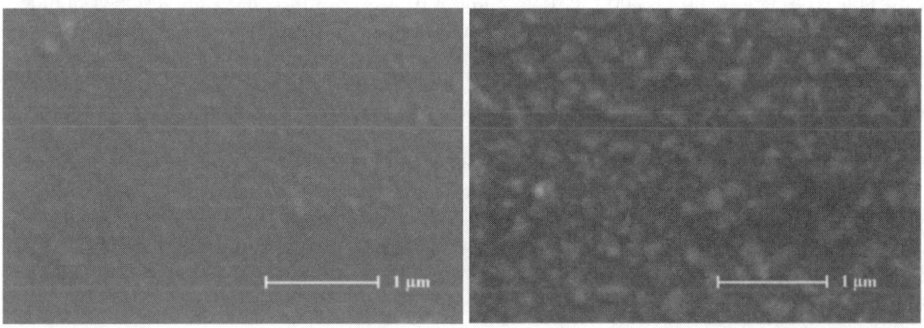

Figure 3. 30000x SEM of Fe_2O_3 thin films prepared at a) 350 °C and b) 450 °C.

Optical Properties

Figure 4 (i) shows the transparency changes of the Fe_2O_3 thin films obtained at different deposition temperatures. It can be observed that the films are darker when the deposition temperature is increased. Figure 4 (ii) shows the transmission spectra of Fe_2O_3 films for wavelengths from 190 to 1100 nm. This figure shows that the average transmission decreases when the substrate temperature increases. The films deposited between 350 and 450 °C show a transmission increase after 600 nm, whereas the films deposited at 200 and 250 °C exhibit a maximum at around 700 nm; these results are in good agreement with earlier studies for amorphous Fe_2O_3 films [10]. Additionally, the transmission spectra show the wavelength cut-offs at around 310 nm, and are transparent in the visible region; this behaviour is typical of ferric oxides such as maghemite [9]. The obtained spectra confirm the presence of maghemite and match with the XRD results.

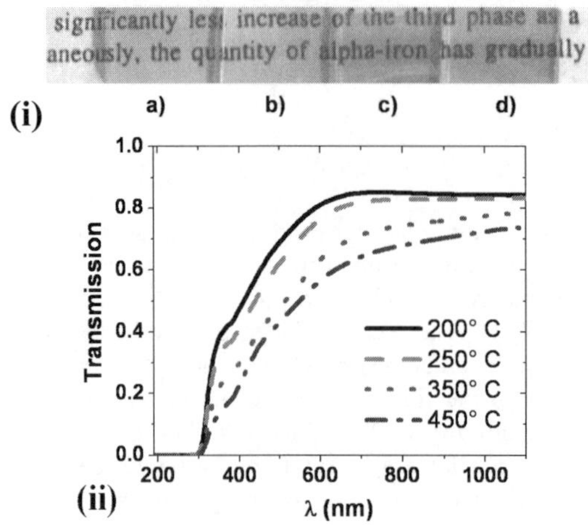

Figure 4. (i) Changes in the transparency of the Fe_2O_3 thin films deposited at a) 200, b) 250, c) 350 and d) 450 °C. (ii) Transmission spectra of Fe_2O_3 deposited at different temperatures.

Conclusions

The thermodynamic analysis performed in this work was a very important tool before the spraying deposition. It facilitates the selection of both the most adequate precursor and deposition temperature for the Fe_2O_3 thin film synthesis. Both the thermodynamic analysis and the experimental work, suggest that the most acceptable precursor is the $FeCl_3$ salt. In comparison with the traditional (pneumatic) technique, the ultrasonic spray pyrolysis technique allows the deposition of both crystalline and amorphous Fe_2O_3 thin films with a high quality, in a fast and reproducible way. Additionally, an increase in the substrate temperature promotes a decrease in the optical transmission, rougher surfaces, and the development of a crystalline phase. It can be concluded that amorphous films present the best optical behaviour (higher transmission) for the design of thin film solar filters.

Acknowledgment

This work was supported by the Cinvestav Research Initiative and the Conacyt. The authors thank Dr. Dagoberto Vazquez from ITS for his assistance in XRD measurements.

Reference

[1] B. Chertok, B. A. Moffat, A. E. David, F. Yu, C. Bergemann, B. D. Ross, V. C. Yang, Biomater, 29 (2008) 487-496.
[2] J. Chavez-Galan, R. Almanza, Solar Ener. 81 (2007) 13-19.
[3] C. D. Mee, E. D. Daniel, F. Jorgensen, Magnetic Recording Technology, 2a Ed. New York: McGraw-Hill, 1995.
[4] W. Eerenstein, T. T. M. Palstra, S. S. Saxena, T. Hibma, Phys. Rev. Lett. 88, 247204 (2002).
[5] B. Orel, M. Macek, F. Svegl, Thin Solid Films 246 (1994) 131.
[6] J. Sarradin, M. Ribes, A. Guessous, K. Elkacemi, Solid State Ionics 112 (1998) 35–40.
[7] C. Jorand Sartoretti, M. Ulmann, B.D. Alexander, J. Augustynski, A. Weidenkaff, Chem. Phys. Lett 376 (2003) 194–200.
[8] H. L. Hartnagel, A. L. Dawar, A. K. Jain, C. Jagadish, "Semiconducting Transparent Thin Films", Ed. Institute of Physics Publishing, Bristol Philadelphia, (1995).
[9] L. Dghoughi, B. Elidrissi, C. Bernède, M. Addou, M. A. Lamrani, M. Regragui ,H. Erguig, Appl. Surf. Sci. 253 (2006) 1823–1829.
[10] A. A. Akl, Appl. Surf. Sci. 233(2004) 307.

Materials Science Forum Vol. 644 (2010) pp 109-112
© (2010) Trans Tech Publications, Switzerland
doi:10.4028/www.scientific.net/MSF.644.109

Synthesis and Atomic Force Microscopy contact current images of aluminum doped ZnO thin films

N. Muñoz Aguirre[1, a], J. E. Rivera López[1, b], L. Martínez Pérez[2, c] and P. Tamayo Meza[1,d].

[1]Instituto Politécnico Nacional, Sección de Estudios de Posgrado e Investigación, Escuela Superior de Ingeniería Mecánica y Eléctrica-UA. Av. Granjas, N° 682, Colonia Santa Catarina. Del. Azcapotzalco, CP. 02550, México, DF. México.

[2]Unidad Profesional Interdisciplinaria en Ingeniería y Tecnologías Avanzadas del Instituto Politécnico Nacional, Av. IPN No. 2580, Col. Barrio La Laguna Ticomán, C.P. 07340, México D.F. México.

[a]nmag804@avantel.net, [b]eduardo_rivera1@hotmail.com, [c]buhomartin@yahoo.com, [d]ptamayom@ipn.mx.

Keywords: ZnO thin films, Atomic Force Microscopy, c-AFM.

Abstract Aluminum doped ZnO thin films were synthesized by the water-mist assisted spray pyrolysis technique. The structural characterization by means of X-Ray diffraction measurements is reported. By means of Atomic Force Microscopy, the superficial electrical characteristics of the thin films are studied. Specifically, contact current images are shown and discussed. It is important to emphasize that in spite of no voltage is applied to the Atomic Force Microscopy contact conductive tip, current images are getting.

Introduction

The synthesis of transparent ZnO thin films with low electrical resistivity is of widely importance for their potential application as transparent conductors in solar cells [1]. For determining the electrical properties of thin films, the four points (or Van der Pauw) method is generally used [1]. Such method is described as follows; four electric contacts over the surface of the thin film are localized. A voltage is applied on two of them and a current is reading at the two other ones. From the resulting I-V curve, the macro electrical resistivity of the sample is obtained. On the last years, a new method for determining local electrical properties of materials has been established [2,3] based on Atomic Force Microscopy. When a standard contact tip is metal coated and bias with respect to the sample, an electrical current is produced [2-4] while the surface is scanning forming a current image. Such technique is called conductive Atomic Force Microscopy (c-AFM) [3]. The principal contribution of this method is the electrical characterization at the nano-scale. In this work, using the c-AFM method, contact current images of the surface of the aluminum doped ZnO thin films are discussed.

Experimental

Aluminum doped ZnO thin films by the water-mist-assisted spray pyrolysis technique [1] were synthesized. An ultrasonic generator operated at 0.8MHz was used for producing the spray of source solution. A 0.3032 M solution of zinc acetylacetonate (from Sigma Aldrich) dissolved in N,N-dimethylformamide (N,N-DMF, from Mallinckrodt) was prepared and 10 % of molar concentration of acetylacetonate of aluminum was added to this chemical solution. High purity air with flows of approximately 10 l/min was used as the carrier gas of the mist solution to the top of the substrate. The synthesis time was ten minutes. A molten tin bath was the substrate heating. Corning 7059 glasses were used as substrates which were carefully cleaned using a well known cleaning procedure [5]. A stream of water mist was simultaneously supplied to the spray of the source solution. The depositions were carried out at substrate temperatures of 500 ºC. X-ray diffraction (XRD) measurements were made using a D-5000 Siemens Diffractometer. The superficial electrical characterization was performed in the configuration shown in figure 1 using a JSPM-5200 microscope from JEOL Ltd.. The used tip was the OSCM-PT model from Veeco Ltd. with a spring constant in the range of $k = 0.5\text{-}4.4$ N/m. Because of was used a new tip for the measurements, the higher value equal to $k = 4.4$ N/m was taking for measure the tip-sample force.

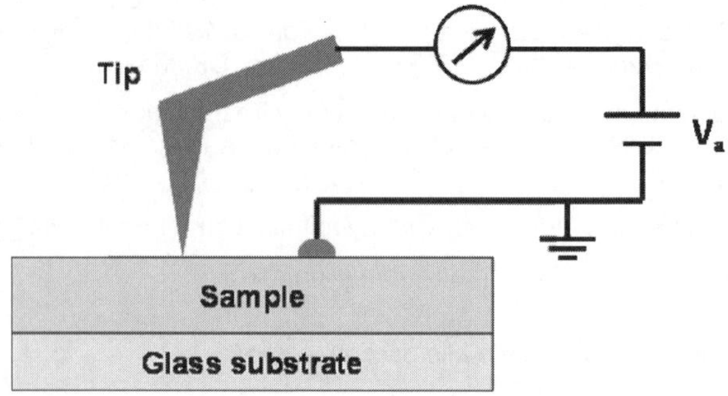

Figure 1. The experimental scheme for getting the current images. V_a is the voltage applied to the conductive tip with respect to a grounded electrical contact localized over the surface of the film. The electrical contact on the surface was made with silver paint.

Results and discussion

The X-Ray diffraction measurements showed the hexagonal structure of the aluminum doped ZnO thin films as can be observed from figure 2. It can be observed the hexagonal phase for both standard and water mist-assisted sprayed samples [1]. A preferential orientation growth is observed for the samples which were deposited with water mist corresponding to the plot with wider line in figure 2. Details of such preferential growth were discussed in reference [6]. From EDS measurements a 5 % of incorporated aluminum [6] was obtained.

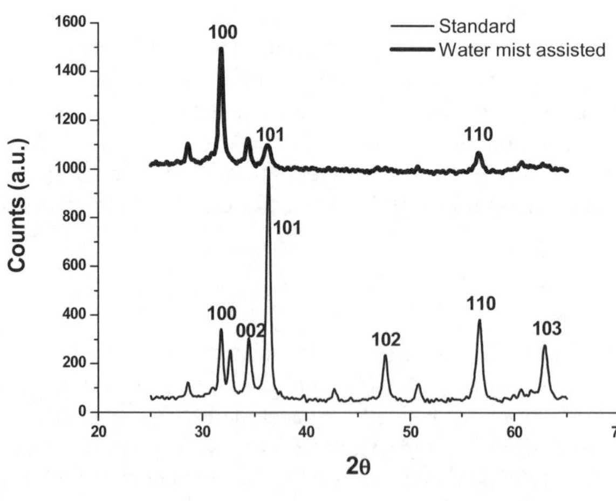

Figure 2. XRD spectra for the aluminum doped ZnO thin films. Current images correspond to the samples deposited with additional water mist.

On the other side, in a nice paper of Zhong Lin Wang and Jinhui Song [7], it is shown that a high piezoelectric current of ZnO nanowires can be produced pressing it with a conductive Atomic Force

Microscope tip. Similar piezoelectric phenomenon in our aluminum doped ZnO thin films is taking place. Pressing the sample with a tip force of the order of 30 nN, an appreciable current up to 10 nA was got without apply a voltage on the tip (*V_a=0)* as can be observed in figure 3. Therefore, if a comparison with the measured electrical resistivity by means of the four point method is wanted, then a lower tip-sample force must be applied. That means the magnitude of reference parameter

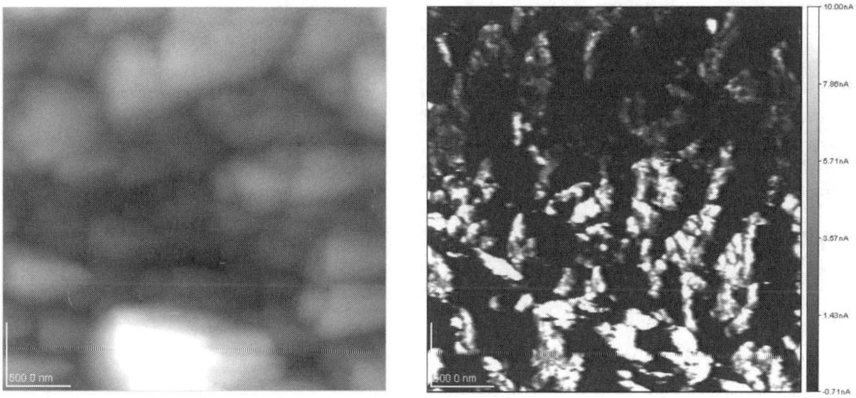

a) b)

Figure 3. Topography a) and their corresponding current b) images of the surface of the aluminum doped ZnO thin films at $V_a=0$. The scanned area is 3×3 μm. The average current measured by the RMS Roughness of the current image is of the order of 2.36 nA.

of the microscope must be increased until the piezoelectric current will be unappreciable. Such situation is shown in figure 4. After the piezoelectric effect of the ZnO thin films was minimized, current images at the same reference magnitude of the last case and applying a voltage of 0.5 V were got as shown in figure 5. It is observed a high current average of 1.36 nA with uniform maxima up to 10 nA.

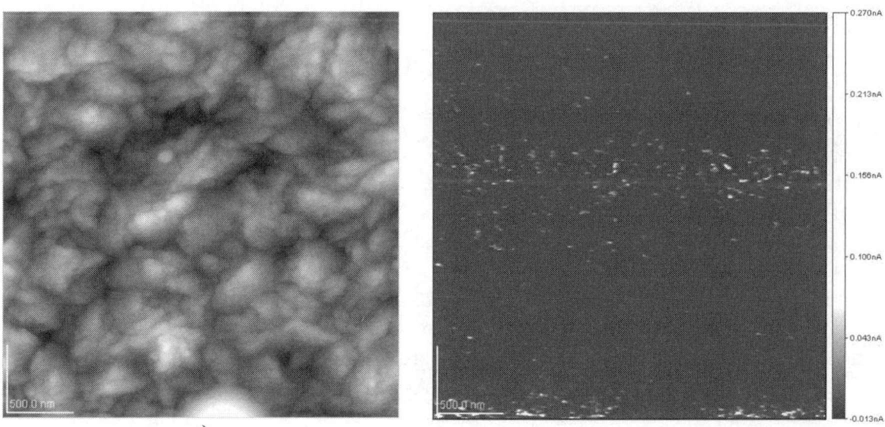

a) b)

Figure 4. Topography a) and their corresponding current b) images of the surface of the aluminum doped ZnO thin films at $V_a=0$ using a tip-sample force of the order of 3 nN. The scanned area is 3×3 μm. The piezoelectric current is very low with the average value of 0.009 nA.

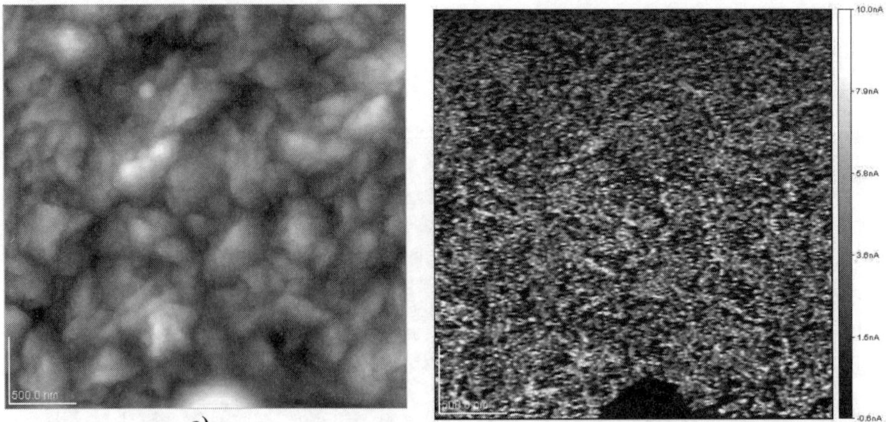

a) b)

Figure 5. Topography a) and their corresponding current b) images of the surface of the films at $V_a=0.5$ V and a tip-sample force of the order of 3 nN. Same scanned area from figure 4.

Summary

From figure 3, it is observed that in spite of no voltage is applied to the AFM contact conductive tip an appreciable current up to 10 nA is got opening the frame of potential applications as that reported in reference [7]. From the figure 5, it is suggested that only after the piezoelectric effect is minimized, the superficial electrical properties of the ZnO thin films can be related with those got by means of the four point method. Also, from the current image of figure 5, a uniform conductivity of the micro-grains can be observed.

References

[1] L. Martínez Pérez, M. Aguilar-Frutis, O. Zelaya-Angel, and N. Muñoz Aguirre, Physica Status Solidi a, **203, No. 10**, (2006) 2411-2417.
[2] Liam S. C. Pingree, Obadiah G. Reid, David S. Ginger, *Adv. Mater.*, **21 No 1**, (2009)19-28.
[3] C. Ionescu-Zanetti, A. Mechler, S. A. Carter, R. Lal, *Adv. Mater.*, **16**, (2004) 385-389.
[4] Ernst Meyer, Scanning Probe Microscopy: The Lab on a Tip. Springer-Verlag Berlin Heidelberg New York (ISBN 3-540-43180-2) (2004).
[5]Micro concentrated cleaning solution, International Products, Burlington, N.J. http://www.ipcol.com/
[6] L. Martínez Pérez, N. Muñoz Aguirre, M. Aguilar Frutis, and O. Zelaya Angel. The aluminium doping of ZnO thin films by means of the water mist assisted spray pyrolysis methodTHIN, Poster sessions at the XVIII-International Materials Research Congress (2009).
[7] Zhong Lin Wang and Jinhui Song, *Science* 14 April 2006:**Vol. 312. no. 5771**, pp. 242 – 246.

Materials Science Forum Vol. 644 (2010) pp 113-116
© (2010) Trans Tech Publications, Switzerland
doi:10.4028/www.scientific.net/MSF.644.113

Structural and morphological properties of $Hf_xZr_{1-x}O_2$ thin films prepared by Pechini route

L.A. García-Cerda[1, 2, a], B.A. Puente-Urbina[1, b], M.A. Quevedo-López[2, c], B.E. Gnade[2,d], L. Baldenegro-Pérez[2, e] H.N. Alshareef[2,3,f], M.A. Hernández-Landaverde[4, g]

[1]Centro de Investigación en Química Aplicada. Departamento de Materiales Avanzados. Blvd. Enrique Reyna Hermosillo #140, C.P. 25253, Saltillo, Coahuila. México.

[2]University of Texas at Dallas, Department of Materials Science and Engineering. Richardson, TX, 75080 USA.

[3]Department of Materials Science, King Abdullah University of Science and Technology Jeddah, Saudi Arabia.

[4]Centro de Investigación y de Estudios Avanzados del IPN. Libramiento Norponiente # 2000, C.P. 76010, Fracc. Real de Juriquilla, Santiago de Querétaro, Qro. México

[a]lagarcia@ciqa.mx, [b]bapuente@ciqa.mx, [c]mquevedo@utdallas.edu, [d]gnade@utdallas.edu [e]leobaldenegro@yahoo.com, [f]husam.alshareef@kaust.edu.sa, [g]martin@qro.cinvestav.mx

Keywords: thin films; dielectric materials; Pechini route; hafnium oxide; zirconium oxide.

Abstract. In this study, $Hf_xZr_{1-x}O_2$ ($0 < x < 1$) thin films were deposited on silicon wafers using a dip-coating technique and by using a precursor solution prepared by the Pechini route. The effects of annealing temperature on the structure and morphological properties of the proposed films were investigated. $Hf_xZr_{1-x}O_2$ thin films with 1, 3 and 5 layers were annealed in air for 2 h at 600 and 800 °C and the structural and morphological properties studied by X-ray diffraction (XRD) and scanning electron microscopy (SEM). XRD results show that the films have monoclinic and tetragonal structure depending of the Hf and Zr concentration. SEM photographs show that all films consist of nanocrystalline grains with sizes in the range of 6 - 13 nm. The total film thickness is about 90 nm.

Introduction

Hafnium oxide (HfO_2) and zirconium oxide (ZrO_2) thin films have attracted great interest, because of their large number of potential applications in optical coatings and microelectronics. Both have been considered as very interesting candidates for replacing conventional SiO_2 as insulator in electronic applications [1]. Polycrystalline HfO_2 and ZrO_2 exist in three polymorphs: monoclinic, tetragonal, and cubic. They have similar chemical structures and are completely miscible in each other and they can form solid solutions, $Hf_xZr_{1-x}O_2$, over the complete concentration range ($0 < x < 1$).

Thin films of HfO_2 and ZrO_2 have been fabricated by various techniques such as sputtering [2], electron-beam evaporation [3], metalorganic chemical vapor deposition (MOCVD) [4], atomic layer deposition (ALD) [5], and solgel [6]. These techniques are well developed and lead to satisfactory results. However, they are expensive and require special equipment, as vacuum chamber, expensive chemicals precursors and appropriate targets. Wet chemical processes are alternative routes, the Pechini route is a sol-gel method attractive for preparing thin films because of its easy control, low cost and low processing temperature [7]. In this method, the metallic salt is dissolved in a mixture of citric acid (CA) and ethyleneglycol (EG) giving rise to a polyester network containing the metallic ions homogeneously distributed. The polymeric solution is applied in the substrate and, generally, the metal oxide thin film is obtained by the calcination at adequate temperatures. The Pechini method has been used to prepare oxides films [8], but the preparation of $Hf_xZr_{1-x}O_2$ thin films has not yet been reported.

Herein, we report the preparation of precursor material by Pechini route and their application for thin film deposition of $Hf_xZr_{1-x}O_2$ by dip coating. The structural and morphological properties of the thin films were studied by X-ray diffraction (XRD) and scanning electron microscopy (SEM).

Experimental

The chemicals used in this study were analytical grade purity supplied by Aldrich Co.: hafnium chloride ($HfCl_4$), zirconium acetate hydroxide ($C_{12}H_{36}O_{24}Zr_3$), citric acid ($C_6H_8O_7.H_2O$) and ethylene glycol ($C_2H_6O_2$) and distilled water. All reagents were used as received. The precursor solution was prepared as follows: first, citric acid was dissolved in distilled water and then the appropriate amounts of $HfCl_4$ and $ZrC_2H_4O_2$ (dictated by the value of x) were added to the citric acid solution and stirred until a clear and homogeneous solution was obtained. Finally, ethylene glycol was added and the resulting solution was stirred for 30 minutes.

For the preparation of $Hf_xZr_{1-x}O_2$ thin films, the precursor solution was poured into a vessel. The cleaned Si(100) substrates (2.5 x 2.5 cm) were immersed in this solution and withdrawn at the rate of 300 mm/min. The deposit was then dried at 100 °C for 15 minutes and finally annealed at 600 °C and 800 °C for 2 h. In order to increase the film thickness, the above deposition process was repeated for three and five times.

X-ray diffraction (XRD) patterns of the thin films were obtained using a Rigaku Ultima III diffractometer with grazing angle mode featuring Cu Kα radiation (λ = 1.5418 Å), a 1° beam incidence angle was used with a step time of 1 s, step size of 0.02°, and tube conditions of 40 kV and 44 mA. The surface and cross-sectional morphologies of the films were examined using a Zeiss Supra-40 Scanning Electron Microscope (SEM); the images were acquired using the secondary electrons detector.

Results and discussion

Fig. 1 shows the XRD patterns of the $Hf_{0.7}Zr_{0.3}O_2$ thin films with different layers annealed for 2 h in air at 600 °C (a) and 800 °C (b), respectively. The result shows that all peaks correspond to the monoclinic phase [9], with the two strongest peaks corresponding to the (-111) and (111) reflections. For the thin film with 1 layer, these peaks are small and increased its intensity when the layers were increased. As the annealing temperature was increased, the diffraction peaks observed were sharper with higher intensity (Fig. 1b), consistent with larger crystal size.

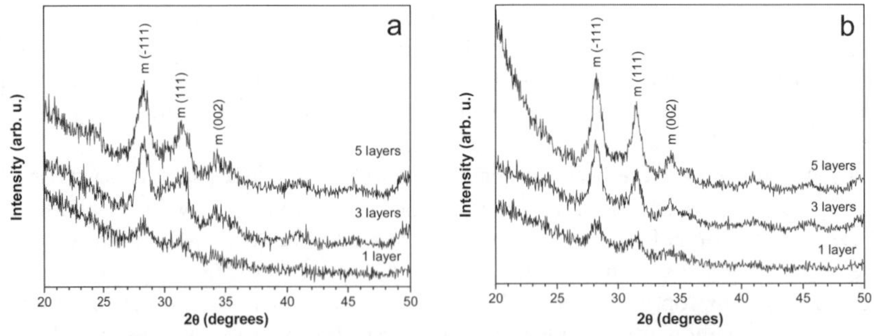

Fig. 1. X-ray diffraction patterns of $Hf_{0.7}Zr_{0.3}O_2$ thin films with different layers and annealed at 600 °C (a) and 800 °C (b).

The influence of the Zr content in the XRD patterns of the thin films (5 layers) is shown in the Fig. 2. For the film of HfO_2, the observed peaks correspond to monoclinic phase (-111 and 111 reflections) but when the Zr content was increased, we observe a decrease in the peaks corresponding to monoclinic phase and an increase in the tetragonal phase of the ZrO_2 (111 and 200 reflections) [10]. The Table 1 summarizes the phases found for all the thin films prepared.

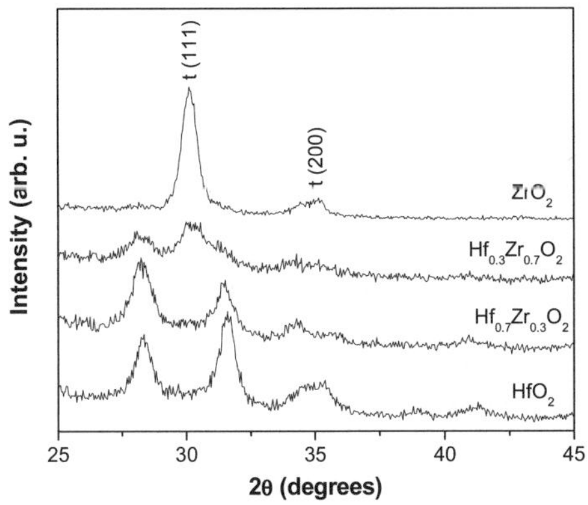

Fig. 2. XRD patterns $Hf_xZr_{1-x}O_2$ as a function of Zr content.

Sample	Crystalline structure	
	600 °C	800 °C
HfO_2	Monoclinic	Monoclinic
$Hf_{0.7}Zr_{0.3}O_2$	Monoclinic	Monoclinic
$Hf_{0.3}Zr_{0.7}O_2$	Monoclinic + Tetragonal	Monoclinic + Tetragonal
ZrO_2	Tetragonal	Tetragonal

Table I. Crystalline structure of the $Hf_xZr_{1-x}O_2$ thin films.

SEM micrographs of the surface morphology of the $Hf_{0.7}Zr_{0.3}O_2$ thin films with 5 layers and annealed at 600 °C (a) and 800 °C (b) are shown in Fig. 3. The micrographs show that the films are crack-free with uniform distribution of $Hf_{0.7}Zr_{0.3}O_2$ nanoparticles. Film thickness of the $Hf_{0.7}Zr_{0.3}O_2$ sample annealed at 800 °C, was determined from cross-sectional SEM micrographs, as shown in Fig. 3c. The SEM image shows that the thickness of the film is approximately 90 nm, similar thicknesses were determined for the rest of the samples. The Table II shows the particle size calculated from the SEM micrographs. The sizes of the nanocrystalline particles are in the range of 6 to 13 nm, when the number of layers and the temperature were increased, the particle size was increased.

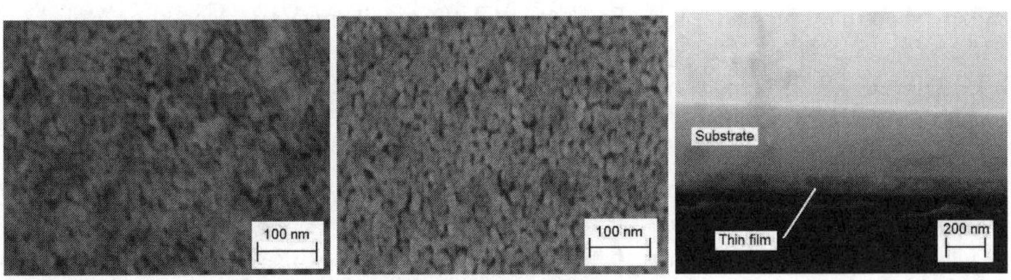

Fig. 3. SEM micrographs showing surface of the $Hf_{0.7}Zr_{0.3}O_2$ annealed at 600 °C (a), 800 °C (b) and cross section (c).

Sample	Particle size (nm)					
	600 °C			800 °C		
Layer	1	3	5	1	3	5
HfO_2	---	10.0	11.5	---	12.5	12.6
$Hf_{0.7}Zr_{0.3}O_2$	---	6.0	6.5	---	10.5	11.5
$Hf_{0.3}Zr_{0.7}O_2$	---	6.1	6.1	---	10.7	12.9
ZrO_2	---	7.4	8	---	13	12.6

Table II. Particle size of the $Hf_xZr_{1-x}O_2$ thin films.

Summary

$Hf_xZr_{1-x}O_2$ thin films have been synthesized by Pechini route. The thin films were annealed in air at 600 and 800 °C. XRD results confirm that the films have monoclinic and/or tetragonal phase depending of the Zr content. The $Hf_xZr_{1-x}O_2$ thin films are crack-free, with homogeneously distributed nanoparticles with sizes in the range of 6 to 13 nm. Currently, additional studies are carrying for determining if these thin films can be used as dielectric insulator in electronic applications.

Acknowledgements

L.A. García-Cerda thanks CONACYT-México for the economical support for the sabbatical leave in the University of Texas at Dallas (UTD). Also thanks to UTD for providing access to the XRD and SEM equipments.

References

[1] G.D. Wilk, R.M. Wallace, J.M. Anthony, J. Appl. Phys. Vol. 89 (2001), p. 5243.
[2] T.V. Perevalov, V.A. Gritsenkoa, S.B. Erenburg, A.M. Badalyan, H. Wong, C.W. Kim, J. Appl. Phys. Vol. 101 (2007), p. 053704-1.
[3] I.P. Studenyak, M. Kranjcec, O.T. Nahusko, O.M. Borets, Thin Solid Films Vol 476 (2005), p.137.
[4] U. Patil, R. Thomas, A. Milanov, R. Bhakta, P. Ehrhart, R. Waser, R. Becker, H. Werner Becker, M. Winter, K. Merz, R.A. Fischer and A. Devi, Chem. Vap. Deposition Vol. 12 (2006), p. 172.
[5] D. H. Triyoso, R. I. Hegde, J. K. Schaeffer, R. Gregory, X.D. Wang, M. Canonico, D. Roan, E.A. Hebert, K. Kim, J. Jiang, R. Rai, V. Kaushik, and S.B. Samavedam, J. Vac. Sci. Technol. B Vol. 25(2007), p. 845.
[6] T. Kidchob, L. Malfatti, F. Serra, P. Falcaro, S. Enzo, P. Innocenzi, J. Sol-Gel. Sci. Techn. Vol. 42 (2007), p. 89.
[7] J. Liu, W.J. Weng, W.H. Ding, K. Cheng, P.Y. Du, G. Shen, G.R. Han, Surf. Coating Technol. Vol. 198 (2005), p. 274.
[8] M. Pocuca, G. Brankovic, Z. Brankovic, D. Vasiljevic-Radovic, D. Poleti, J. Eur. Ceram. Soc. Vol. 27 (2007), p. 1083.
[9] Joint Committee on Powder Diffraction Standards (JCPDS); Card No. 043-1017.
[10] Joint Committee on Powder Diffraction Standards (JCPDS); Card No. 050-1089.

Materials Science Forum Vol. 644 (2010) pp 117-122
© (2010) Trans Tech Publications, Switzerland
doi:10.4028/www.scientific.net/MSF.644.117

Extended crystallographic defects in gallium nitride

Yadira Arroyo Rojas Dasilva[1,a], Pierre Ruterana[1,b], Lise Lahourcade[2], Eva Monroy[2], Gilles Nataf[3]

[1]CIMAP, UMR 6252 CNRS-ENSICAEN-CEA-UCBN, 6, Boulevard Maréchal Juin, 14050 Caen Cedex 4 France
[2]INAC/SP2M/PSC, 17 rue des Martyrs, 38054 Grenoble Cedex 9 France
[3]CRHEA UPR 10 CNRS, 1, Rue Bernard Gregory, 06560, Valbonne, France
[a]yadira.arroyo@ensicaen.fr, [b]pierre.ruterana@ensicaen.fr

Keywords: GaN, defects, dislocations, stacking faults, polar, non-polar and semi-polar, TEM.

Abstract. In the fabrication GaN-based devices, several growth orientations are currently under investigation in order to exploit material properties which are inaccessible using layers grown along the *c*-axis. However, such procedures rely on foreign substrates with large misfits. Therefore, complex crystallographic defects form in the epitaxial layers and have been the subject of extensive studies. They include threading dislocations and stacking faults, which can be within basal or prismatic planes. Out of the *c*-axis, depending on the growth orientation, the glide planes of perfect dislocations may be no longer available, complicating the relaxation processes.

Introduction

For almost the last two decades, III-nitride semiconductors (AlN, GaN, and InN) have been subject of extensive investigations due to their many actual as well as potential applications for optoelectronics and electronics [1, 2]. C-plane is the most common orientation for the growth of III-nitride films [1]. Transmission electron microscopy (TEM) applied to *c*-oriented nitride layers pointed out the occurrence of several types of extended defects, such as threading dislocations [3], stacking faults (basal and prismatic) [4-7], inversion domains (IDs) [8] and grain boundaries (GBs) [9]. These defects have deleterious effects on the performance of commercial devices, particularly in terms of lifetime and reliability [10]. For the last decade, tremendous improvement of the material quality has been attained in various ways. First, the optimization of the nucleation process and buffer layers allowed the growth of unipolar nitride layers and the elimination of the stacking faults in thick templates [3]. More recently, the use of the epitaxial lateral overgrowth (ELO) allowed reducing the threading dislocation density from the initial 10^{10} cm^{-2} to the low 1×10^{6} cm^{-2} range inside the active layers [11, 12]. These improvements led to the production of blue laser diodes with lifetimes above 10^{4} h [1].

In spite of the progress in material growth, *c*-layers suffer from intrinsic limitations due to polarization effects [13], which induce strong electric fields in heterostructures and contribute to reduce the radiative recombination efficiency of nitride optoelectronic devices [14]. One approach to avoid these effects is to grow layers along directions perpendicular (non-polar) or inclined (semi-polar) to the [0001] direction. Non-polar layers are grown on *a*-$(11\bar{2}0)$ and *m*-$(10\bar{1}0)$ orientations. In such layers, the [0001] polar axis is parallel to the heterostructure interfaces, so that the fabricated devices become free of polarization effects. Unfortunately, the attained internal efficiency of non-polar devices remains low due to the high defect density, mainly threading dislocations ($\sim1\times10^{10}$ cm^{-2} for *a*-GaN and 4×10^{9} cm^{-2} for *m*-GaN) and basal stacking faults ($\sim1\times10^{6}$ cm^{-1} for *a*-GaN and 1×10^{5} cm^{-1} for *m*-GaN) which form along the growth direction and cross all the active device areas [15,16]. In semi-polar layers, the [0001] axis lies at an angle of the heterostructure interfaces, and the polarization effects, are at least attenuated. Several semi-polar orientations are currently under investigation, such as $(11\bar{2}2)$ [17,18], $(11\bar{2}6)$ [19] or $(10\bar{1}3)$ [17] . In the following, we shall review and discuss the extended defects in polar, non-polar and semi-polar nitride layers as observed by conventional TEM and high-resolution microscopy. Their stability and formation mechanisms will be discussed.

Defects in Wurtzite Structure

In wurtzite nitride semiconductors, the most common extended defects are dislocations and stacking faults. The Burgers vectors for perfect dislocations in wurtzite structure, shown in Figure 1, are ±*c* (***b*** = <0001>), ±*a* (***b*** = $1/3<11\bar{2}0>$) and their sum *c*±*a* (***b*** = $1/3<11\bar{2}3>$). In order to decrease the dislocation energy, perfect dislocations may be dissociated into the partial dislocations (PDs). In the wurtzite structure, the Frank and Shockley PDs have $1/6<20\bar{2}3>$ and $1/3<10\bar{1}0>$ Burgers vector, respectively. Regarding basal stacking faults (BSFs), in wurtzite structure we find two types: intrinsic (I) and extrinsic (E) BSFs (Figure 1). Two intrinsic BSFs can be distinguished; they are called I_1 and I_2. The I_1-BSFs corresponds to one violation of the hexagonal stacking changing from the normal sequence to AaBbAaBbCcBbCcBb, it has a displacement vector $R = 1/6<20\bar{2}3>$. The I_2 stacking sequence and the displacement vector are AaBbAaBbCcAaCcAaCc and $R = 1/3<10\bar{1}0>$, respectively. The E-BSF has a displacement vector $R = 1/2<0001>$, corresponding to an insertion of three cubic sequences in the hexagonal stacking: AaBbAaBbCcAaBbAaBb.

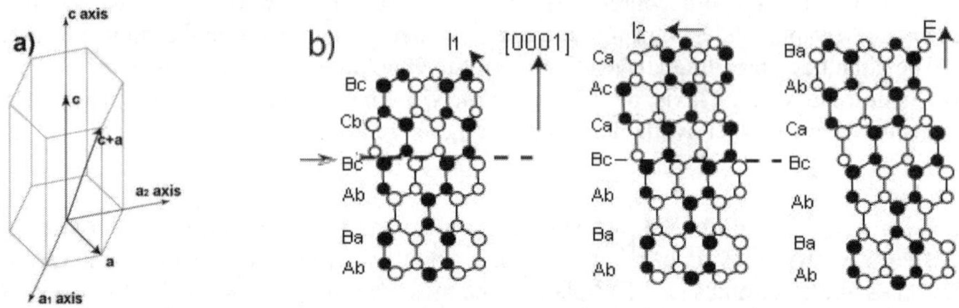

Fig. 1. Defects in the wurtzite structure: a) Burgers vector for dislocations (a, c, a + c); b) Intrinsic and extrinsic BSFs (I1, I2, E) stacking sequences, the arrow, (horizontal) indicates starting point

In III-nitrides, I_1–BSFs have low formation energy [20, 5]; they are largely extended along the (0001) planes and often only one of the terminating partial dislocations is observed. Another type of stacking fault whose atomic structure has been investigated in nitride layers is the prismatic stacking fault (PSF). Such defects have been shown to form inside $(11\bar{2}0)$ planes during the layer growth from surface steps [21], and they exhibit two atomic structures corresponding to $1/2<10\bar{1}1>$ and $1/6<20\bar{2}3>$ displacement vectors, respectively [6]. Moreover, due to the lack of a symmetry centre along the 6-fold axis of the wurtzite structure, the growth along the [0001] direction may lead to the formation of areas of different polarity, which may cross the whole epitaxial layer, or form closed domains [22]. In these materials where the ionicity is higher than in conventional III-V semiconductors such as GaAs, theoretical *ab initio* investigations reported that the boundaries of such domains would need to be reconstructed in one atomic structure for stability [23]. However, observations showed that their configuration exhibited the two possible structures depending on the growth conditions [24].

Polar Nitrides

The epitaxial relationship of *c*-plane GaN and *c*-sapphire is $(0001)_{GaN} \| (0001)_{sapphire}$ and $[01\bar{1}0]_{GaN} \| [11\bar{2}0]_{sapphire}$. In this configuration, there is a large lattice mismatch of 16% in the basal plane and the growth of GaN on sapphire results in the formation of numerous defects. Without special care for the epitaxial growth, the threading dislocation (TDs) density inside layers is of the order of $10^{10}\ cm^{-2}$. These dislocations have Burgers vectors $b = 1/3<11\bar{2}0>$, $b = [0001]$ and $b = 1/3<11\bar{2}3>$. TEM investigations in high-quality *c*-GaN layers grown by MOCVD have shown that the majority of TDs have an edge character with a Burgers vector equal to $1/3<11\bar{2}0>$ and a [0001] dislocation line [3]. Screw TDs have small percentage of the total defects density (1-10%) in *c*-GaN layers grown using MOVPE [21]. Using molecular beam epitaxial (MBE), the percentage of screw TDs increases up to 20% overall density [25]. The mixed-type dislocations percentage varies from 2% using MBE to 68% using MOVPE [25,26]. The formation mechanism of threading dislocations is has been under debate for a long time. Taking into account the large mismatch between sapphire and III-nitrides, the first idea that came out was that the growth of these materials should follow a mosaic process [27], and the thick layers result from coalescence of previously relaxed grains [28] which may have undergone various degrees of twist and/or tilt. This coalescence process was therefore due to give rise to grain boundary dislocations [9]. However, other processes may also be playing an important role. Many BSFs have been reported close and parallel to the GaN/sapphire interface [3, 7, 29], whereas the density of these BSFs tends to decrease markedly along the layer thickness. For this reason, some authors suggested that the disappearance of BSFs in the buffer layers came from their reaction and could give rise to the formation of the threading dislocations [30]. Indeed, a substantial reduction of the density of TDs in nitride layers has been possible due to the use of the ELO technique, which is mainly the growth of a nitride layer through micron-size windows opened in a SiO_2 mask deposited on a nitride buffer layer. The lateral overgrowth is favoured above the mask and the dislocations are forced to bend in the basal plane; their density is thus reduced along the *c*-axis. This reduction does not take place within the coalesced areas, where there are reactions, and in the centre of the openings, where dislocation lines mostly stay vertical, respectively [11,12].

Non-polar and Semi-polar Nitrides

The investigation of non-polar and semi-polar nitride layers is more recent: It has only been going on for the last 6-7 years [31]. In these orientations, the basal planes of wurtzite structure which are parallel to the stacking sequences of the system are either normal (*a, m*) or make a finite non zero angle (semipolar) with the substrate surface. An important characteristic of the microstructure in such materials is the SFs density, which is high in comparison with c-grown nitrides, does not decrease from the low temperature buffers layers to the surface.

***a*-plane GaN.** The epitaxial relationship for a-GaN on r-sapphire are $[10\bar{1}0]_{GaN} \| [11\bar{2}0]_{sapphire}$ and $[0001]_{GaN} \| [\bar{1}101]_{sapphire}$. The corresponding lattice mismatches in these directions are $\approx 1.1\%$ and 16%, respectively, therefore, the strain relation may be rather complex due to this non symmetric configuration. At the interface, this lattice mismatch is accommodated by the formation of misfit dislocations. The *a*-GaN/*r*-sapphire interface has steps as is shown in Figure 2a. In spite of the small lattice mismatch along the *m*-direction, a high defect density has been observed, particularly threading dislocations and basal stacking faults lying in the basal plane.

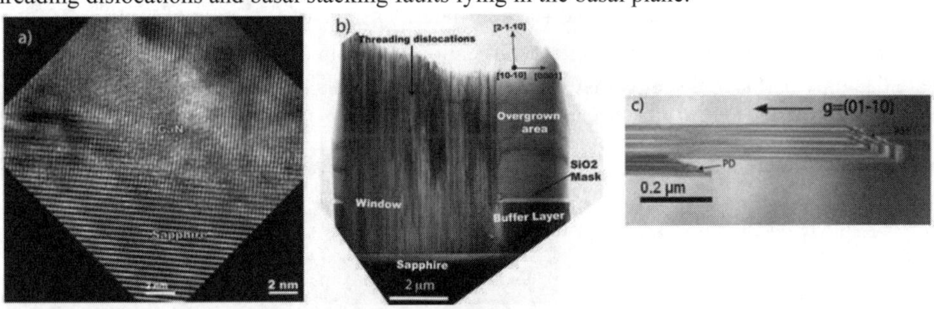

Fig. 2. Micrographs of a-GaN layers: a) a-GaN/r-sapphire interface viewed along the $[10\bar{1}0]_{GaN}$ zone axis, b) $g = 11\bar{2}0$ weak beam bright field micrograph with vertical the threading dislocations (vertical arrow), and c) basal stacking faults ending with a partial dislocation (D) and prismatic stacking faults

Threading dislocations and stacking faults that are created at the *a*-GaN/*r*-sapphire interface propagated along the whole layer towards the top surface (Figure 2b), as shown in the part marked window. The dislocation density is of the order of $1x10^{10}\ cm^{-2}$ [15]. Their Burgers vectors are $1/3[\bar{2}110]$ and $1/3<11\bar{2}3>$ and, like in c-GaN layers, their line direction are parallel to the growth direction. The stacking fault density is $1x10^6\ cm^{-1}$- they are basal stacking faults of the type I_1 and I_2. In this instance also ELO has been used in attempts to reducing the defects density: subsequently, it was possible to attain dislocation and stacking fault densities of $3x10^8\ cm^{-2}$ and $6x10^4\ cm^{-1}$, respectively [15]. Prismatic stacking faults with a displacement vector ½$[1\bar{1}01]$ are also observed in *a*-GaN layers. In this case, PSFs may originate from the coalescence of islands [15].

***m*-plane GaN.** There is less reported information on *m*-GaN layers than on *a*-GaN layers, probably due to the lack of available substrates, and/or additional growth difficulties. The more common substrates for *m*-GaN layers are SiC [32] and γ-$LiAlO_2$ (LAO) [17, 32]. On LAO, the epitaxial relationship is $[11\bar{2}0]_{GaN} \| [001]_{LAO}$. High TD and BSF densities have also been reported in m-GaN layers. Subsequent ELO implementation has allowed decreasing the TD density from $4x10^9\ cm^{-2}$ to $6x10^8\ cm^{-2}$ [17]. In fact, the TDs have different character than in polar layers: they can be of edge type with $b = 1/3[11\bar{2}0]$ and mixed with $b = 1/3 < 11\bar{2}3 >$ [33]. In such layers, BSFs have been identified as I_1 type [16], using ELO, their density could be decreased from $1x10^5\ cm^{-1}$ in standard growth down to $3x10^3\ cm^{-1}$.

Semi-polar GaN. As can be noticed above, even using the ELO technique the density of extended defects remains rather high in non-polar nitride layers. As a result, even if these materials are free of polarization effects, their use in efficient devices is still out of hand. More recently, there has been a rising interest for the investigation of semi-polar GaN layers along various orientations, such as $(11\bar{2}2)$ [17,18], $(11\bar{2}6)$ [19], or $(10\bar{1}\bar{3})$ [17] growth planes. In this case, the polarization is reduced depending on inclination of the *c*-axis to the interface. One of the most efficient semi-polar orientation for GaN is $(11\bar{2}2)$-GaN, where the c-axis lies at 58° off the interfaces. It has been recently shown that a substantial increase of device efficiency may be attained on this crystallographic orientation [34]. In an extensive investigation by MBE, it has been shown that this orientation is dominant on top of *m*-sapphire [35], with in-plane epitaxial orientation $[11\bar{2}3]_{GaN} \| [0001]_{sapphire}$ and $[1\bar{1}00]_{III\text{-}N} \| [11\bar{2}0]_{sapphire}$; a small fraction of $\{10\bar{1}3\}$-domains was pointed out mainly for GaN when it is deposited directly on sapphire. However the optimization of the growth conditions and the systematic use of an AlN buffer layer allow obtaining exclusively the $(11\bar{2}2)$-GaN on *m*-sapphire [36]. There also remains a high defect density in $(11\bar{2}2)$-GaN layers. Although one report has mentioned dislocation and stacking fault densities of the order of $2x10^{10}\ cm^{-2}$ and $2x10^5\ cm^{-1}$ respectively [17], a complete identification is still under way and preliminary results indicate that only partial dislocation may be present. In this instance, the defects lie inside the basal plane at $\approx 58°$ from the $(11\bar{2}2)$ growth plane (Fig. 3).

Fig. 3. Weak beam micrograph taken in g/5g = $10\bar{1}0$ conditions: some partial dislocations (PD) are marked and the BSFs are visible in c-planes. The growth direction is indicated by the white arrow

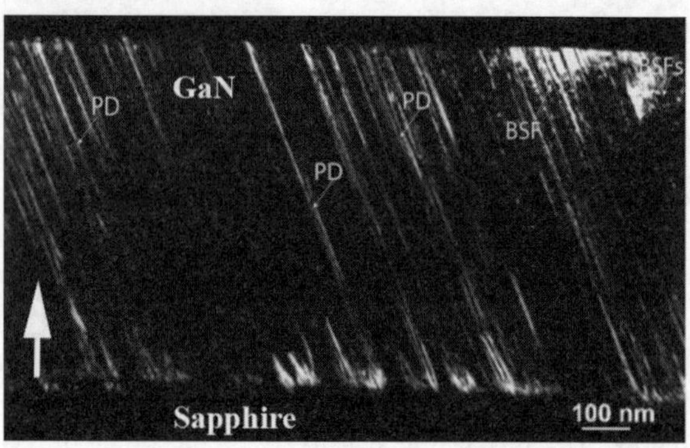

In $(11\bar{2}6)$-GaN layers, the $(11\bar{2}6)$ plane is parallel to the $(1\bar{1}02)$ plane of sapphire. High dislocation and stacking fault densities are found. The dislocation density is $8x10^9\ cm^{-2}$. Higher stacking fault density is found close to the interface than at the top surface [19]. They are of the type I_1 and I_2 bounded by $1/6<20\bar{2}3>$ and $1/3<10\bar{1}0>$ partial dislocations. In $(10\bar{1}\bar{3})$-GaN, the epitaxial relationships arc $[30\bar{3}\bar{2}]_{GaN} \parallel [12\bar{1}0]_{sapphire}$ and $[1\bar{2}10]_{GaN} \parallel [0001]_{sapphire}$. Basal stacking faults are observed with a density of $2x10^5\ cm^{-1}$ and threading dislocations with a density of $9x10^8\ cm^{-2}$ [16].

Discussion and Conclusion

In this short review of extended defects in the nitrides, we focus on the study of GaN, since the types of defects that form are similar in the different wurtzite nitride compounds, although the particular morphology may change [37-39]. It is now accepted that the defects can have various atomic configurations depending on the layer growth conditions (IDBs [8, 23, 24], PSFs [4, 6, 40], dislocations [25, 41, 42]). In fact, more than one configuration of a defect may even coexist through interaction with other defects [43]. The growth conditions and the substrate surface state appear also to govern the formation of some extended defects: on the (0001)-sapphire and (111)-Si surfaces, it has been shown that the formation of IDs may be connected with the residual misfits along the c-axis which is introduced by steps [44, 45]. On SiC, the surface steps are at the origin of the formation of the PSFs in GaN [46]. As for the threading dislocations, it is obvious that the strain relaxation at the substrate surface plays an important role in their formation [27, 28, 47] as well as the various defect reactions that take place inside the buffer layer [30]. In non-polar and semi-polar layers, the geometrical relationships at the growth surface are more complex than on the top of c-substrates. The symmetry of the surface is reduced in comparison to the pseudo-6-fold of the (0001)-sapphire. Moreover, the glide planes of the wurtzite structure dislocations, which coincide with the prismatic planes, do no longer contain the growth axis. As it can be noticed in the example given above, defects in non-polar and semi-polar layers do not propagated along the growth direction, but are forced to lie in the c-planes. Therefore, it may be concluded that in these wurtzite semiconductors layers, the observed crystallographic defects are formed during growth and their atomic structures appear to be mostly driven by the kinetics.

Acknowledgement: The authors acknowledge the support of the National Research Agency under contract N° ANR-08-BLAN-0298-01 and the European Commission under the Grant agreement N°: PITN-GA-2008-213238, Initial training network RAINBOW of the 7 RTD Framework.

References

[1] S. Nakamura, S. Pearton and G. Fasol: *The blue Laser Diodes*. Springer, Germany, 2000.

[2] H. Markoc. *Nitrides semiconductors and devices*. Springer, Germany 2001.

[3] X. H. Wu, L. M. Brown, D. Kapolnek, S. Keller, S. P. DenBaars, and J. S. Speck: J. Appl. Phys. Vol 80 (1996), p. 3228.

[4] P. Vermaut, P. Ruterana and G. Nouet: Philos. Mag. A. Vol 76 (1997), p. 1215.

[5] P. Ruterana, B. Barbaray, A. Bere, P. Vermaut, A. Hairie, E. Paumier, G. Nouet, A. Salvador, A. Botchkarev, and H. Morkoc: Phys. Rev. Vol 59 (1999), p. 15917.

[6] P. Vermaut, G. Nouet, and P. Ruterana: Appl. Phys. Lett. Vol 74 (1999), p. 694.

[7] V. Potin, P Ruterana and G Nouet: J. Phys.: Condens. Matter. Vol 12 (2000), p. 10301.

[8] V. Potin, P. Ruterana, and G. Nouet: J. Appl. Phys. Vol. 82 (1997), p. 2176.

[9] V. Potin, P. Ruterana, G. Nouet, R. C. Pond and H. Morkoc: Phys. Rev. B. Vol 61 (2000), p. 5587.

[10] J. W. P. Hsu, M. J. Manfra, R. J. Molnar, B. Heying and J. S. Speck: Appl. Phys. Lett. Vol 81 (2002), p. 79.

[11] T. S. Zhelevaa, O. H. Nam, W. M. Ashmawic, J. D. Griffin, R. F. Davis: J. Cryst. Growth. Vol 222 (2001), p. 706.

[12] B. Beaumont, Ph. Vennegues, and P. Gibart: Phys. Stat. Sol. (b). Vol 227 (2001), p.1.

[13] F. Bernardini, V. Fiorentini and D. Vanderbilt: Phys. Rev. B. Vol. 56 (1997), p. R10024.

[14] D. Du and D. J. Srolovitz: Acta Mater. Vol 52 (2004), p. 3355.

[15] Y. A. R. Dasilva: Characterization of a-plane grown GaN on sapphire substrates by electron microscopy. PhD thesis, EPFL, No.4193, 2009.
[16] B.A. Haskell, T. J. Baker, M. B. McLaurin, F. Wu, T. Fini, S. P. DenBaars, J. S. Speck and S. Nakamura: Appl. Phys. Lett. Vol.86 (2005), p.111917.
[17] T. J. Baker, B.A. Haskell, F. Wu, J. S. Speck and S. Nakamura: Jpn. J. Appl. Phys: Part 2. Vol 45 (2006), p. L154.
[18] T. Gühne, Z. Bougrioua, P. Vennéguès, M. Leroux and M. Albrecht : J. Appl. Phys. Vol 101 (2007), p. 113101.
[19] L. Zhou, R. Chandrasekaran, T.D. Moustakas , D. J. Smith: J. Cryst. Growth. Vol 3010 (2008), p. 2981.
[20] C. Stampfl and C.G. Van de Walle: Phys . Rev. B. Vol 57 (1998), R15052.
[21] P. Vermaut, P. Ruterana and G. Nouet. Philos. Mag A. Vol 75 (1997), p. 239.
[22] V. Potin, G. Nouet, P. Ruterana: Philos. Mag. A. Vol 79 (1999), p. 2899.
[23] J. E. Northrup, J. Neugebauer and L. T. Romano: Phys. Rev. Lett. Vol 77 (1996), p. 103.
[24] V. Potin, G. Nouet, P. Ruterana: Appl. Phys. Lett. Vol 74 (1999), p. 947.
[25] Y. Xin, S. J. Pennycook, N. D. Browning, P. D. Nellist, S. Sivananthan, F. Omnes, B. Beaumont, J. P. Faurie and P. Gibart: Appl. Phys. Lett. Vol 72 (1998), p.2680.
[26] N. Grandjean, J. Massies, P. Vennegues, M. Leroux, F. Demangeot, M. Renucci, and J. Frandon : J. Appl. Phys. Vol 86 (1998), p. 1379.
[27] F.A. Ponce: MRS Bulletin. Vol 22 (1997), P. 51.
[28] Th. Kehagias, P. Komninou, G. Nouet, P. Ruterana and Th. Karakostas: Phys. Rev. B. Vol 64 (2001), p. 195329.
[29] H. K. Cho, J. Y. Lee, K. S. Kim, G. M. Yang, J. H. Song and P. W. Yu: J. Appl. Phys. Vol 89 (2001), p.2617.
[30] V. Narayanan, K. Lorenz, W. Kim, and S. Mahajan: Appl. Phys. Lett. Vol 74 (2001), p. 1544.
[31] M. D. Craven, S. H. Lim, F. Wu, J. S. Speck, and S. P. DenBaars: Appl. Phys. Lett. Vol 81 (2002), p. 469.
[32] T. Kawashima, T. Nagai, D. Iida, A. Miura,Y. Okadome, Y. Tsuchiya, M. Iwaya, S. Kamiyama, H. Amano, I. Akasaki: Phys. Stat. Sol. Vol 244 (2007), p. 1848.
[33] I. Lo, C.H.Hsieh, Y. L. Chen, W. Y. Pang, Y. C. Hsu, J. C. Chiang, M. C. Chou, J. K. Tsai, and D. M. Schaadt: Appl. Phys. Lett. Vol 92 (2008), p. 202106.
[34] L. Lahourcade, P.K. Kandaswamy, J. Renard, P.Ruterana, H. Machhadani, M. Tchernycheva, F.H. Julien, B. Gayral, and E. Monroy: Appl. Phys. Lett. 93 (2008), p. 111906.
[35] L. Lahourcade, E. Bellet-Amalric, E. Monroy, M. Abouzaid and P. Ruterana: Appl. Phys. Lett. Vol 90 (2007), p. 131909.
[36] L. Lahourcade, J. Renard, B. Gayral, E. Monroy, M. P. Chauvat and P. Ruterana: J. Appl. Phys. Vol 103 (2008), p. 093514.
[37] V. Potin, P. Vermaut, P. Ruterana, and G. Nouet: J. Elec. Mat.Vol 27 (1998), p. 266.
[38] J. Jasinski, Z. Liliental-Weber, Q. S. Paduano and D. W. Weyburne: Appl. Phys. Lett. Vol 83 (2003), p. 2811.
[39] J. Jasinski, Z. Liliental-Weber, H. Lu, and J. W. Schaff: Appl. Phys. Lett. 85 (2004), p.233.
[40] J.E. Northrup: Appl.Phys.Lett.Vol.72(1998), p. 2316.
[41] J. Chen, P. Ruterana and G. Nouet: Phys. Rev. B Vol 67 (2003), p. 205210.
[42] J.E. Northrup:Appl.Phys.Lett.,Vol.78(2001), p. 2288.
[43] A. M. Sanchez, G. Nouet, P. Ruterana, F. J. Pacheco, S. I. Molina and R. Garcia : Appl. Phys. Lett. Vol 79 (2001), p. 3588.
[44] P. Ruterana, V. Potin, B. Barbaray and G. Nouet: Philos. Mag: A Vol 80 (2000), p. 937.
[45] A.M. Sanchez, P. Ruterana, S.I. Molina, F.J. Pacheco, R. Garcia: Phys. Stat. Sol. B234(2002), p. 935.
[46] P. Vermaut, P. Ruterana, G. Nouet and H. Markoc: Philos. Mag. A Vol 75 (1997), p. 239.
[47] Y.B. Kwon, J.H. Je, P. Ruterana and G. Nouet:J. Vac. Sci. Technol. A23 (2005), p. 1588.

Materials Science Forum Vol. 644 (2010) pp 123-127
© (2010) Trans Tech Publications, Switzerland
doi:10.4028/www.scientific.net/MSF.644.123

Structural Characterization of Poly(sodium 4-styrene sulfonate)/CdS Semiconductor Nanoparticle Composites

Matthew P. Espe, [1, a] Saida Y. Ortiz-Colon, [1,b] Arturo Ponce [2,c] and Ronald F. Ziolo [2,d]

[1]Knight Chemical Laboratories, Department of Chemistry, University of Akron, Akron, OH USA 44325-3601

[2] Centro de Investigación de Química Aplicada, Saltillo, Coahuila, México 25253

[a]espe@uakron.edu, [b]syo1@uakron.edu, [c]aponce@ciqa.mx, [d]rziolo@cs.com

Keywords: Nanoparticles, composites, polymer, solid-state NMR.

Abstract. The properties of polymer/semiconducting nanoparticle (NP) composites-materials used in hybrid, bulk-heterojunction photovoltaic materials-are dependent on the interaction of the NPs and polymer. Composite films of water soluble polymers and CdS NPs have been produced both by synthesizing the NPs within the polymer matrix and by adding the CdS NPs, containing a capping agent, to the polymer. The composites have been characterized by microscopy as well as ^{1}H, ^{13}C and ^{113}Cd solid-state NMR. When synthesizing the NPs within the polymer, the polymer matrix plays a role in the cadmium sulfide NP nucleation, growth and structure. In the blended system, the dominant interaction between the glycerol capping agent and sulfonated polymer is observed to be hydrogen bonding.

Introduction

Cadmium chalcogenide (CdX; X = S, Se, Te) semiconducting nanoparticles (NP) are currently of significant interest as the properties of the NP are intermediate to those of molecular and bulk materials, and these materials have a large and diverse number of areas of realized and potential application [1,2]. A number of studies have appeared where these semiconductor NPs are combined with polymers for use in solid-state applications. In most cases the NPs are added to the polymers to enhance the NP processibility, where thin films can be readily fabricated. The dispersion of the NP within the polymer is also an effective method to restrain particle aggregation. The interface between the NP and polymer is critical is defining the properties of these materials. In the case of hybrid materials for photovoltaic applications, the interface controls the rate of forward and reverse electron transfer rates, ultimately defining the power conversion efficiency. The present studies were undertaken to determine the impact of the polymer on the NP size and structure and the geometry of the polymer/NP interface.

Experimental

Synthesis of CdS Nanoparticles in Poly(sodium 4-styrene sulfonate) (CdS/PSS). One gram of poly(sodium 4-styrene sulfonate) (Aldrich, ~70,000 MW)) was slowly dissolved in dionized water and mixed with 1.5 grams of $Cd(NO_3)_2{\cdot}4H_2O$ in water. The solution was placed in cellulose dialysis tube (Sigma) and left stirring overnight in deionized water. The dialysis tube was rinsed several times until no Cd ion was detected. The solution was again dialyzed against water for two 24 hr periods, and the solution was mixed with an alkaline solution of excess $Na_2S{\cdot}9H_20$. The solution was placed in a dialysis tube and left stirring overnight in deionized water. The dialysis tube was rinsed several times until no S ion was detected in solution. The solution was again dialyzed against water for two 24 hr periods, and then placed in an open Petri dish to air dry.

Composites of CdS nanoparticle and Poly(sodium 4-styrene sulfonate) (CdS-TG/PSS). CdS nanoparticles capped with 1-thioglyceral (TG) were synthesized as reported previously [3]. The CdS nanoparticle solution was dialyzed against DI water three times over a 24 hr period. To the solution was added poly(sodium 4-styrene sulfonate) (Scientific Polymer Products, Mw ~ 70,000) and the solution was stirred for 2 hrs. The solution was then placed in a 15 cm diameter Petri dish and allowed to dry under ambient conditions.

HRTEM. Transmission electron microscopy (TEM) using selected area electron diffraction (SAED) and high resolution (HRTEM) were carried out in a FEI-TITAN 80-300 kV field emission gun microscope, which has a symmetrical condenser-objective lens type S-TWIN (C_s = 1.25 mm). The HRTEM images acquired near the Scherzer focus and SAED patterns were registered in a CCD camera and subsequently analyzed using Digital Micrograph software.

NMR. Solid-state ^{113}Cd and ^{13}C solid-state NMR spectra were acquired on Varian Unity/Plus (4.7 T, 9.4 T) and Varian DirectDrive (11.7T) instruments. All ^{113}Cd spectra were collected using magic-angle spinning (MAS) with a sample spinning speed of 5 kHz (4.7 T), and the chemical shifts were referenced to 1.0 M $Cd(ClO_4)_2$ (0 ppm). Direct polarization (DP) data were collected using a 30 s recycle delay, and cross polarization (CP) experiments were performed using a mixing time of 3ms, recycle delay of 4s, with proton decoupling. The ^{13}C spectra were collected under CP/ MAS with a sample spinning speed of 12 kHz (9.4 T), mixing time of 3ms, recycle delay of 4s and proton decoupling. Solid state ^{1}H NMR (11.6 T) spectra were obtained using MAS with a sample spinning speed of 13 kHz, a 90° pulse length of 4.0 μs and a 5 s recycle delay.

Results and Discussion

CdS nanoparticles synthesized within the Poly(sodium 4-styrene sulfonate) matrix. Based on the TEM data the CdS NPs synthesized in PSS by ion-exchange and precipitation have diameters in the range of 3-7 nm and are well dispersed in the polymer matrix. A HRTEM micrograph, Fig. 1, shows several particles near the film edge, and the selected area electron diffraction data reveals that the NP have adopted the cubic lattice type. The ^{113}Cd solid-state NMR (SSNMR) spectrum from the CdS NPs is much broader than the signal from bulk CdS, Fig. 2. This broadening results from the structural disorder of the NP surface, which accounts for ~ 30% of the Cd sites in the NPs. The ^{113}Cd CP/MAS NMR spectrum from CdS/PSS, Fig. 2, arises from only the surface Cd sites and the additional broadening of the peak is consistent with increased disorder at the surface.

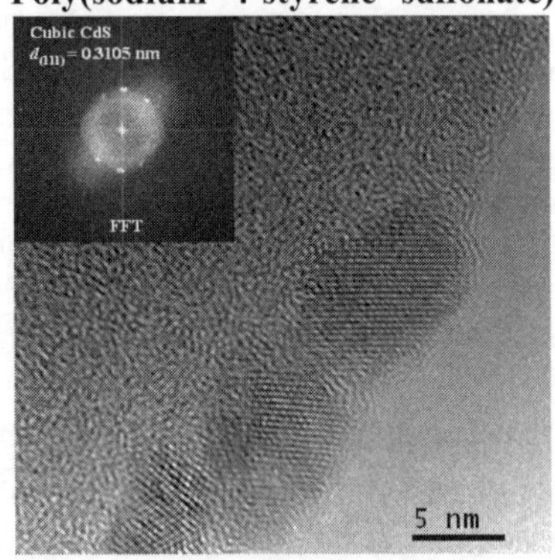

Figure 1. HRTEM image and FFT from CdS in the CdS/PSS film.

Typical linewidths in the ^{113}Cd NMR spectra from CdS and CdSe NPs are 45-65 ppm, synthesized either with or without a capping agent, and the linewidth is dependent on the NP synthetic procedure [4]. NPs synthesized in solution in the presence of a capping agent have linewidths nearer 45 ppm [5,6]. In contrast, CdS NPs formed by simply combining cadmium and sulfide ion solutions results in NPs with a diameter of ~4nm and a ^{113}Cd NMR linewidth of ~65 ppm [4]. In the current work there is no capping agent present during the NP synthesis, except for the solubilized PSS; in this case, the linewidth is nearer the lower end at 45 ppm. These results suggest that the polymer matrix plays a role in cadmium sulfide NP nucleation and growth. However, the mechanism by which the polymer impacts the structure of the NPs, formed in the polymer's presence, is currently not understood.

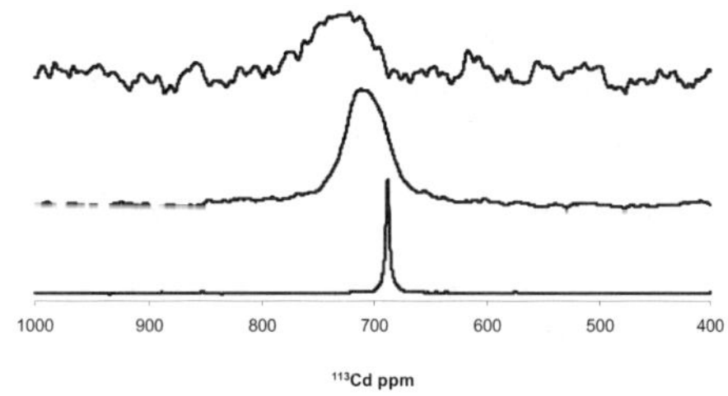

Figure 2. ^{113}Cd DP/MAS NMR spectra from bulk CdS (bottom) and CdS/PSS (middle). The top spectrum is the CP/MAS spectrum from CdS/PSS.

Composite of CdS nanoparticle and Poly(sodium 4-styrene sulfonate). The synthesized CdS NPs have a mono-layer coating of 1-thioglyceral (TG) as the capping agent. Absorption and HRTEM studies indicate that the coated NPs (CdS-TG) have a size of 5±0.5 nm. Interaction between the TG and the polymer in the composite is evident from the ^{1}H SSNMR spectra in Fig. 3. The spectrum from TG attached to CdS has a peak at 7 ppm arising from the -OH protons and the spectrum from PSS has a peak at ~4ppm from adsorbed water. The water in the polymer will be located in the vicinity of the sulfonate groups. The peak in the NMR spectrum from the alcohol proton of TG is shifted and broadened in the composite, indicating a change in the electronic structure at the –OH groups and a reduction in the motion of these groups, respectively. The restricted motion is consistent with an interaction between the TG and polymer, with the most likely interaction involving the formation of hydrogen bonds between the TG (-OH) and the SO_3^- groups of the polymer. The peak from the absorbed water is modified in a similar manner upon composite formation, indicating that the water is also located at the interface and has reduced dynamics.

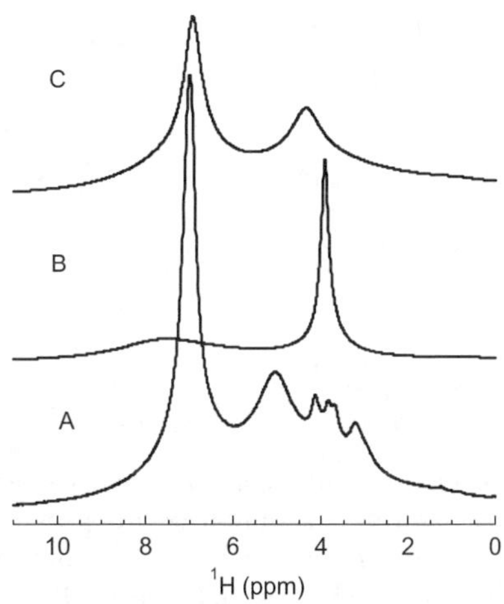

Figure 3. ^{1}H SSNMR spectra from A) CdS-TG, B) PSS and C) CdS-TG/PSS composite.

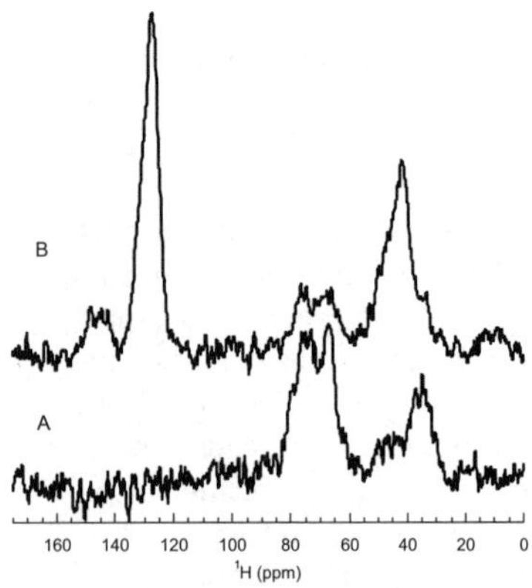

Figure 4. ^{13}C SSNMR spectra from A) CdS-TG and B) CdS-TG/PSS.

The details of the molecular dynamics of the polymer backbone, polymer side chain and TG have been probed by using ^{13}C SSNMR, Fig.4. The resolution achieved at a field strength of 11.7 T allows for the identification of peaks from each species present: PSS backbone, 41 ppm; PSS phenyl group, 127, 144 and 147 ppm; TG, 67 and 75 ppm. The relaxation value $T_1\rho(^1H)$, a value sensitive to molecular motions in the kHz regime, has been measured for PSS and CdS-TG separately and for the CdS-TG/PSS composite, Table 1. The lack of change in the relaxation times between the pure polymer and the composite means that the dynamics of the organic portion of the polymer are little changed upon formation of the composite. In contrast, the relaxation values for TG show significant change, which means the frequency and/or magnitude of the molecular dynamics of this species has changed. If the frequency/magnitude of the motion of TG has increased upon composite formation then one expects to see a narrowing of the peaks in the NMR data, especially in the ^1H spectrum. As the opposite is observed, the motion of the TG group is restricted in the presence of the polymer.

Table 1. $T_1\rho(^1H)$ values for the CdS-TG/PSS composite

Species	^{13}C Chemical Shift	PSS	CdS-TG	CdS-TG/**PSS**	**CdS-TG**/PSS
PSS-phenyl	127 ppm	9.5 ms		8.5 ms	
PSS-phenyl	144 ppm	23.6 ms		29 ms	
PSS-phenyl	147 ppm	13 ms		11.3 ms	
TG	67 ppm		5.6 ms		2.5 ms
TG	75 ppm		7.5 ms		2.7 ms

Model. The most likely intermolecular interaction between TG and PSS is hydrogen bonding between the sulfonate group of PSS and the glycerol moiety of TG. The hydrogen bonding will reduce the motion of TG and produce the smaller values of $T1\rho(^1H)$ observed. The interface will also contain several water molecules with limited mobility, as well as a sodium cation from the polymer. Using crystal structures located in the Cambridge Crystallagraphic Data Base, a model for the interface structure has been developed and is shown in Fig. 5. The water molecules are likely to hydrogen bond with the glycerol and sulfonate groups and coordinate the sodium ion through their oxygens. In this model the organic portion of the polymer chains at the interface are far enough removed from the TG that their dynamics are only slightly altered by the presence of CdS-TG and we observe only a small change in their $T_1\rho(^1H)$ values.

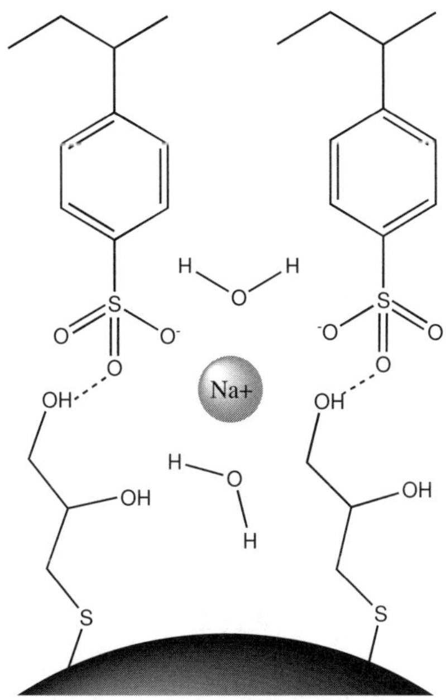

Figure 5. Model of the interface in CdS-TG/PSS film

Conclusion

Composite films consisting of PSS and CdS nanoparticles, either bare or with a capping agent, have been generated and characterized. Synthesis of CdS NP within PSS produces well-dispersed particles, however the dispersion in diameter is larger than for solution synthetic techniques. A result not reported previously is that in the presence of PSS the NPs produced are structurally more homogeneous than in the absence of the polymer. When a capping agent is present, such as thioglyceral, it may interact with the polymer. For the system with PSS, the most important interaction is hydrogen bonding. These studies are currently being extended to composites containing water soluble conjugated polymers.

References

[1] Burda, C.; Chen, X.; Narayanan, R.; El-Sayed, M.A. *Chemical Reviews,* 2005, 105(4), 1025-102.

[2] Alivisatos, A.P. *J. Phys. Chem.,* 1996**,** 100(31), 13226-13239.

[3] Martínez-Castañón, G.A; Sánchez-Loredo, M.G.; Martínez-Mendoza, J.R.; Ruiz, F; *J. Mater. Online*, 1, 1-7, 2005.

[4] Ladizhansky, V.; Hodes G.; Vega, S. *J. Phys. Chem. B,* 1998, 102, 8505

[5] Berrettine, M.G.; Braun, G.; Hu J.G.; Strouse, G.F. *J. Am. Chem. Soc.*, 2004, 126, 7063.

[6] Ratcliffe, C.I.; Yu, K.; Ripmeester, J.A.; Zaman, M.B.; Badarau, C.; S. Singh, S. *Phys. Chem. Chem. Phys.,* 2006, 8, 3510

Materials Science Forum Vol. 644 (2010) pp 129-133
© (2010) Trans Tech Publications, Switzerland
doi:10.4028/www.scientific.net/MSF.644.129

Electron and atomic force microscopy of electrochromic WO_3 and molybdenum doped WO_3 thin films deposited by pulsed spray pyrolysis

Dwight R. Acosta[a], Jesús M. Ortega, Carlos R. Magaña

Instituto de Física, Universidad Nacional Autónoma de México
A.P. 20364, México D .F., México
[a]dacosta@fisica.unam.mx

Abstract. WO_3/FTO/Glass and WO_3:Mo/FTO/Glass thin films were deposited using the pulsed spray pyrolysis technique. The Molybdenum doped thin films were deposited using 2,4,6,8 and 10 % at . concentration in the starting solution. The influence of molybdenum concentration on the structural, electrical, optical and electrochromic properties of WO_3:Mo thin films have been systematically studied using Scanning Electron Microscopy (SEM), Selected Area Diffraction Patterns (SAED) in Conventional Transmission Electron Microscopy (CTEM), High Resolution Electron Microscopy (HREM) and Atomic Force Microscopy (AFM). The electrochemical behavior of our materials was studied using cyclic voltammetry both to induce electrochromic behavior and to characterize the electric charge injection and extraction processes. It was found that samples with 2% at. of molybdenum concentration in the starting solution, show the best electrochromic behavior with the highest efficiency and durability. The influence of structural details in our films and their evolution with the molybdenum incorporation, for a fixed substrate temperature on the electrochromic properties of our WO_3:Mo thin films has been followed wherever possible.

Introduction

Tungsten trioxide [WO_3] is a wide band gap n-type metal oxide semiconductor, with interesting properties, which make it suitable for a variety of applications when prepared as thin and/or thick film. The distinctive physical and chemical properties of WO_3 make it suitable for technological applications as transparent electrochromic devices, selective catalysts for oxidation and reduction reactions or transparent conducting electrodes. WO_3 has been studied during last 30 years for applications in electrochromic devices [1], [2],[3].Spray pyrolysis involves the application of a fine mist of very small droplets containing the desired reactants onto the hot substrates. The most critical operation of the spray pyrolysis technique is the preparation of uniform and fine droplets and the controlled thermal decomposition of these droplets.

Experimental

WO_3/FTO/Glass and WO_3:Mo/FTO/Glass thin films were deposited using the pulsed spray pyrolysis technique with a device built reported by Mooney et al. [4].The tungsten oxide materials were deposited on FTO/Glass substrate. For the FTO production over soda lime glass, depositions were carried out using stannic chloride ($SnCl_4$) dissolved in ethanol (0.2 M), and as fluorine source ammonium fluoride (NH4F) dissolved in water was used. The films prepared for a F/Sn = 0.5 ratio and a substrate temperature of 450 °C were selected because their specific electrical parameters: the sheet resistances, namely with 7 and 10 (**Ω/cm**). Molybdenum doped tungsten trioxide [WO_3:Mo] thin solid films were obtained from a starting solution containing tungsten (VI) chloride [WCl_6] (99.9 % from Sigma-Aldrich) dissolved in dimethylformamide [$HCON(CH_3)_2$] (99.9 % from Baker) with a molar concentration of 0.15 M, and stirred for 5 min., until a total dissolution was observed. Molybdenum (VI) dichloride dioxide [MoO_2Cl_2] (from Alfa Aesar) was used as the dopant source; the percentage atomic ratio of molybdenum to tungsten in the starting solution, [Mo]/[W] was used as the reference of the doping level in this work. The films were deposited with a systematic variation in the atomic content of Mo in the starting solution: 2, 4, 6, 8, and 10 %.at . concentration. Both a JEOL-5600 LV scanning electron microscope in Secondary Electrons (SI) mode and a JSPM-4210 atomic force microscope (AFM) in tapping mode were used to study morphological and surface details. CTEM and HRTEM observations were carried out in a JEOL STEM FEG 2010 microscope in standard operational modes respectively and structural information was obtained from electron micrographs using the Digital Micrograph software.

Results

In the SEM micrographs figures 1a and 1b) for 2 and 4 at.% of Mo in the starting solution surfaces with regular dense and compact configuration are observed. In figure 1, larger and faceted grains are observed everywhere together with an increase in surface roughness and in the inter- grain porosity.

.

Figures 1a and 1b.SEM micrographs from samples deposited with 2 and 4 at.% of Mo in the starting solution. The increase in grain sizes is easily observed in figure 1b.

In Figures 2a and 2b , AFM micrographs are presented. Well defined grain shapes and grain agglomerates can be observed for each Mo concentration. The evolution of grain sizes, surface roughness and inter-grain space distribution as the Mo concentration in the starting solution is increased, can be followed from the sequence in 2D and 3D configurations.

Electron diffractions studies from WO_3:Mo thin films were carried out from the CTEM micrographs obtained in the Selected Area Electron Diffraction (SAED) mode and from the interplanar distances derived using the computing program of J. Labar [5]. In figure 3 analysis, orthorhombic and monoclinic WO_3 phases were detected. Neither Co oxides nor Co-W mixed oxides were detected in the SAED patterns.

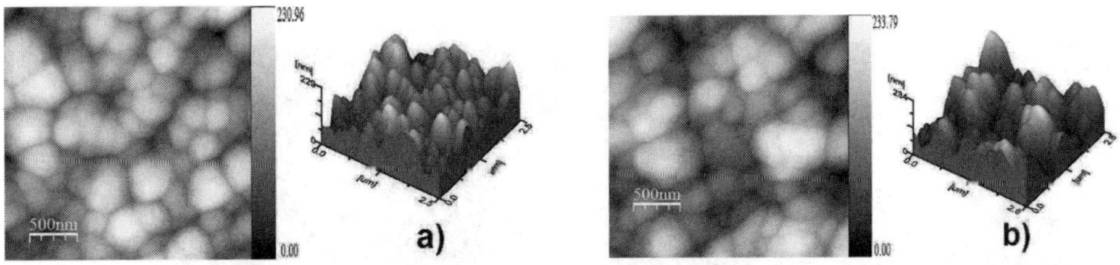

Figures 2a and 2b. AFM micrographs from films deposited with 2 and 4 at% of Mo in the starting solution. The increase in grain sizes and the intergranular spacings together with surface roughness are noticeabe

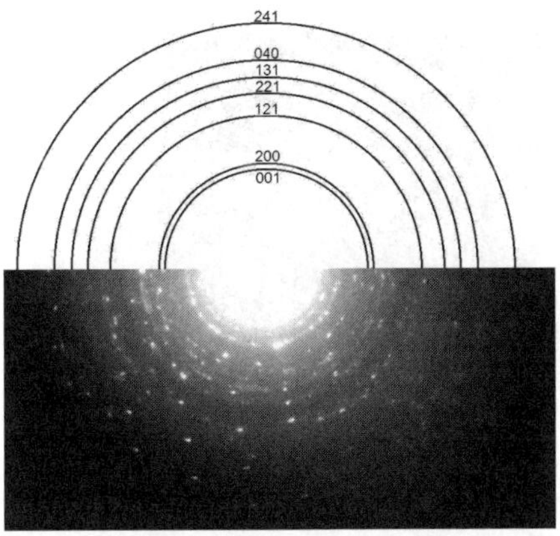

Figure 3. SAED pattern from a polycrystalline WO_3:Mo film with 4% at. Mo concentration.

Table 1

WO_3 PDF #20-1324 Card (Orthorombic)	WO_3 PDF #05-0363 Card (Monoclinic)	SAED RESULTS (This work)
d(Å)	d(Å)	d(Å)
3.853	3.835	3.866
3.692	3.642	3.656
2.525	2.528	2.518
2.173	2.172	2.199
2.018	2.02	2.016
1.873	1.879	1.850

Table 1.Interplanar distances obtained from the SAED pattern displayed in figure 3

In figures 4a and 4b, HREM micrographs of films deposited at 2, 4 at. %of Mo in the starting solution respectively, are presented. In the left side of figure 4a an almost perfect crystalline configuration is observed. The selected squared zone was processed with Fourier transform methods and the corresponding filtered images and spectra are presented in the central and right sides of the figure. The high crystallinity of the nano grains is noticeable and the optical diffraction pattern reveals an orthorhombic crystallographic detail in agreement with the WO_3 distances reported in the PDF # 20-1324 Card. In figure 4b, dispersed nanocrystallites over an amorphous support are visible. Three crystallites without well defined grain borders are displayed in the micrograph. The lattice planes derived after FT analysis from the left side and central part of the image present distances related to WO_3 compounds but the ones in the upper and right side, do not fit with any known tungsten oxide and suggest the possible synthesis of Co-W mixed oxides during the pyrolysis process on glass substrates.

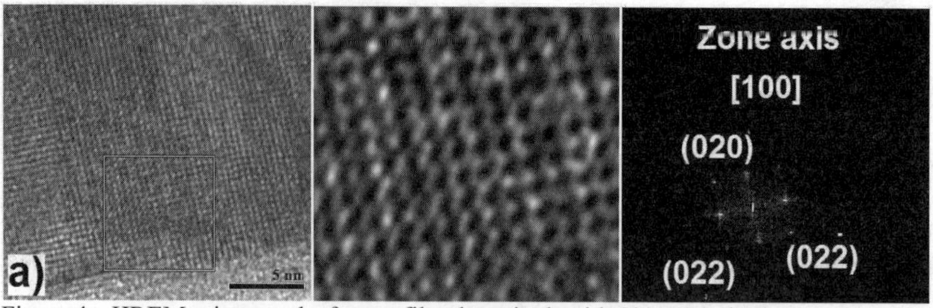

Figure 4a. HREM micrographs from a film deposited with 2% at.Mo in the starting solution.The FT results from processing the squared zone are displayed at the central image and at right side.

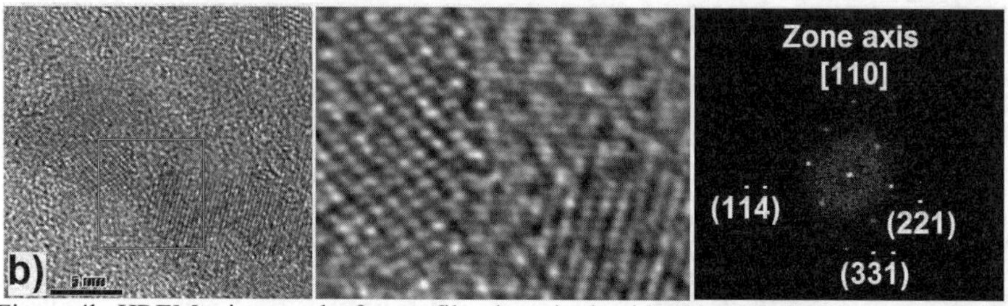

Figure 4b. HREM micrographs from a film deposited with 4% at.Mo in the starting solution.At the upper right side in the inverse FT image, a structure with lattice distances different to the ones of WO_3 can be observed.

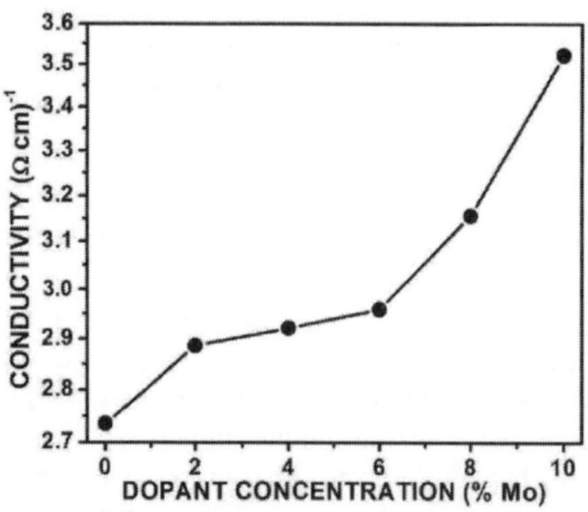

Figure 5. Electrical conductivity behavior as function of Mo concentrations.

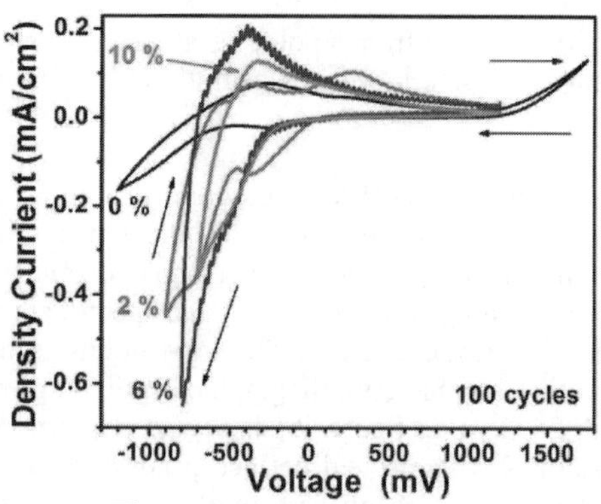

Figure 6.Cyclic Voltammetry of WO_3Mo films

The electrical conductivity was determined using the van der Pauw method and from data shown in figure 5 .The increase of conductivity values is closely related with to the Mo concentration in the starting solution. In figure 6, the electrochemical response under cyclic voltammetry conditions is displayed for films deposited with different starting molybdenum concentrations. The influence of Mo doping concentration on the electrochromic behavior of our films, can be noted in the variations in the intensity and in the position of cathodic and anodic peaks respectively. These parameters depend on structural parameters like size and grain shapes given that they are related to the carrier mean free path, potential contact and charge barriers. amounts and speed of charge extraction and injection processes when a cyclic and variable potential is applied.

Conclusions

Polycrystalline oxides of WO_3:Mo have been obtained successfully using the pulsed spray pyrolysis technique. The addition of molybdenum atoms to the WO_3 during the material synthesis and when the pyrolysis process is going on, over the hot FTO/glass substrate induces important changes in the surface morphology and at minor level induce structural variations at nanometer level, as the molybdenum concentration is increased. From the cyclic voltammetry graphic is apparent that an increase in Mo concentration influences the height of charge injection and extraction peaks. From extended studies carried out for Mo concentration running from 2 to 10 at.%, - not all reported in this short work- the higher electrochromic efficiency and films durability were found for WO_3:Mo thin films prepared with a 2 % of Mo atomic concentration in the starting solution.

Acknowledgments
The authors appreciate and thank the financial support of PIFUT08-129 project from Instituto de Ciencia y Tecnología del Distrito Federal (ICTDF),México D.F., MEXICO

References

1. J. S. E M. Svensoon, C. G. Granqvist, Solar Energy Mater 11 (1984) 29.
2. M. Carpenter and D. A. Corrigan (Eds.), Proc. Symp. Electrocromic Materials. The Electrochem. Soc. 90, (2) (1990).
3. C. M. Lampert, Solar Energy Materials & Solar Cells 76 (2003) 489.
4. J. B. Mooney, S. B. Radding, Ann. Rev. Mater. Sci. 12 (1982) 81
5. J. Labar, Ultramicroscopy, 103 (2005) 237-249

Materials Science Forum Vol. 644 (2010) pp 135-138
© (2010) Trans Tech Publications, Switzerland
doi:10.4028/www.scientific.net/MSF.644.135

Morphologic modification in ABS by the incorporation "*in situ*" of nano-ZnO: Study of the effect on characteristics and final properties

Ramón Díaz de León[1,a], Rebeca Betancourt[1,b], Bertha Puente[1,c], Pablo Acuña[1,d] and Luís Falcón[2,a]

[1]Centro de Investigación en Química Aplicada, Blvd. Enrique Reyna #140, Saltillo, México

[2]Instituto Tecnológico de Saltillo, Blvd. Venustiano Carranza S/N, Saltillo, México

[a]rdeleon@ciqa.mx, [b]rgalindo@ciqa.mx

Keywords: ABS, nanoparticles, zinc oxide, final properties

Abstract. The synthesis of poly(acrylonitrile-butadiene-styrene) (ABS) was carried out using the mass-suspension process, high *cis* polybutadiene (PB), benzoyl peroxide (BPO) as the initiator and modified and unmodified nanoparticles of zinc oxide (nano-ZnO). X-ray diffraction (XRD) and Fourier transform infrared (FTIR) were used to characterize the ZnO nanoparticles. A Universal tensiometer, Brokfield viscosimeter and transmission electronic microscopy (TEM) were used to characterize ABS. The employment of nano-ZnO induced an impressive morphologic modification and transformed the highly occluded salame morphology of poly(styrene-acrylonitrile) (SAN) to a highly dispersed one in its matrix. During the reaction, a decrease in the total production of SAN was observed due to the interaction between radicals produced by the initiator and nano-ZnO, causing a molecular weight increase of SAN. The nano-ZnO also induced a decline in the phase inversion and extended the period of occurrence. ABS without nano-ZnO yielded higher mechanical properties than one without any nanoparticles.

Introduction

Polystyrene (PS) is a thermoplastic polymer that in spite of its low cost, excellent transparency and good processability, offers a low impact resistance, which significantly limits their applications. When this polymer is reinforced with rubber impact polystyrene is obtained (IPS) [1]. Once modified IPS turns into a rigid matrix consisting particles of rubber. Among the most important materials derived from IPS modification is (ABS) whose matrix is formed by SAN. The performance of the ABS is determined by the interaction of different molecular and morphologic parameters in both the continuous SAN phase and the rubber dispersed phase [2]. In spite of the good qualities of ABS such as low cost, ease of processing and especially its excellent properties it is unfortunately being slowly displaced by other polymers with better performance. In this respect, an alternative to increase the ABS performance that would allow it to evolve towards higher performances is trough the reinforcement with inorganic or organic nanoparticles. The employment of nanoparticles as reinforcements in different kind of matrix, would significantly contribute to the improvement of better properties. For instances, silver nanoparticles have the capacity to confer a bactericide character, carbon nanotubes could increase in general the module of systems where they are dispersed into and nanoclays could promote thermal resistance [3]. In the case of nano-ZnO, they could provide an excellent flame resistant character [4]. Some of the most relevant applications of nano-ZnO have been in rubber formulation, to diminish vulcanization time. ABS nanocomposites with reinforced nanoparticles have been carried out by means of mechanical blend, whereas the preparation "*in situ*" is so far unexplored. In this work were carried out the synthesis and characterization of tree ABS´s materials, two of them, ABS nanocomposites with nano-ZnO and study the effect on their morphology and final properties.

Experimental Part

Synthesis and characterization of ZnO nanoparticles

Initially 0.1 M solution of zinc acetate in ethanol was agitated at 80°C for 3 h. Later 0.1 M solution of sodium hydroxide was then added and the mixture was left to react for 2 h. The resulting product was washed, centrifuged and vacuum dried at 80°C for 4 h. Figure 1 shows the XRD pattern and the IR spectrum of the nano-ZnO. It is observed that the material presents the crystalline structure of the wurtzite. The nano-ZnO were modified with an aminosilane compound as reported by Comparán [5].

a) b)

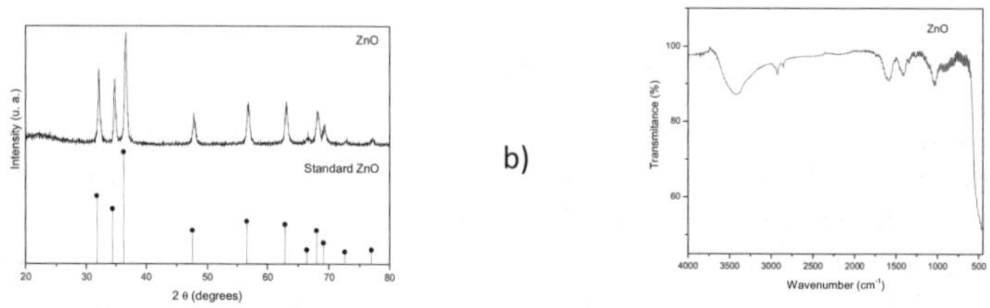

Figure 1. a) XRD pattern and b) Infrared spectrum of the nano-ZnO.

Synthesis and characterization of ABS and the ABS nanocomposites based on ZnO

Table 1 shows the recipes uses to produce the tree ABS´s. The polymerization technique used was the mass-suspension. When ABS was synthetized in the presence of nano-ZnO, the nanoparticles were dispersed initially in 50 mL of an S/AN/PB solution using ultrasound and were later added to the reactor. During the mass stage, samples were taken at different times to evaluate the conversion and Brookfield viscosity in a 10 % w/w solution in toluene at 25°C. The final product was injection molded to evaluate impact strength based on ASTM-256. Tension strength was conducted according to ASTM-D638. The materials also were characterized by TEM to observe the morphology of the ABS´s.

Table 1. Recipes used in the synthesis for the tree ABS´s.

Material	$[PB]_0$	S/AN	$[S/AN]_0$	$[I]_0$	$[AM]_0$	$[CTA]_0$	$[ZnO]_0$
ABS-1	8	22.6/77.4	92	0.5	5	1	0
ABS-2	8	22.6/77.4	92	0.5	5	1	0.5 (unmodified)
ABS-3	8	22.6/77.4	92	0.5	5	1	0.5 (modified)

$[\,]_0$: % w/w initial concentration; PB: polibutadiene; S/AN: styrene/acrylonitrile relation; I: benzoyl peroxide initiator; AM: mineral oil; ATC: chain transfer agent tert-dodecyl mercaptan; ZnO: zinc oxide.

Results and Discussion

Figure 2 shows the behavior of the conversion and viscosity as a function of polymerization time for the three synthetized ABS polymers. It's observed that for ABS-1 (without nano-ZnO) a better production of SAN is reached in a minor reaction time (120 min). The phase inversion crosses a maximum and a minimum in a conversion interval of 14-17 %. The morphology developed in this case corresponds to an elastomeric morphology with multiple occlusions of SAN. When unmodified nano-ZnO are added, ABS-2 results in a conversion decrease with its phase inversion revealing a lower conversion interval of 7-14 % and a slight increase in viscosity. With the incorporation of modified nano-ZnO, ABS-3 behaves in a similar way like ABS-2 but the SAN production is much lower resulting in a phase inversion interval of 7-13 %. The difference in the SAN production between ABS-1and ABS-2 and ABS-3 can be explained due principally to the interaction of primary radicals generated by BPO with nano-ZnO for ABS-2 and ABS-3. The nano-ZnO diminishes the total concentration of radicals in the reaction medium resulted in an increase of their molecular weight (see Table 2). The decrease in the conversion interval from the introduction of the nano-ZnO is fundamentally due to the increase in the molecular weight which in turn causes an increase in the viscosity. On the other hand, the decrease in conversion with unmodified nano-ZnO is also due to the generation of a very unstable system with the production of grafts a seen in Figure 3. But with the addition of modified nano-ZnO, practically eliminates the grafted SAN.

Figure 4 reveal that the morphology of ABS-2 and ABS-3 is not conventional as salami. Instead a rubber phase drastically modified rods shape is observed highly distributed in the SAN matrix. For ABS-2, the unmodified ZnO nanoparticles tend to form agglomerates. For ABS-3the modified ZnO nanoparticles are well dispersed with an impressive morphology, because to the decrease in the production of the grafts as well as in the increase in the size of the nanoparticles. This type of morphology had not been obtained previously before in the synthesis of ABS.

Table 2 reveals that ABS with nano-ZnO, yields higher mechanical properties: impact strength of ABS-2 is up 60 %; Young´s Modulus of ABS increases 15 % and yield strength of ABS-2 and ABS-3 is slowly higher than ABS-1.

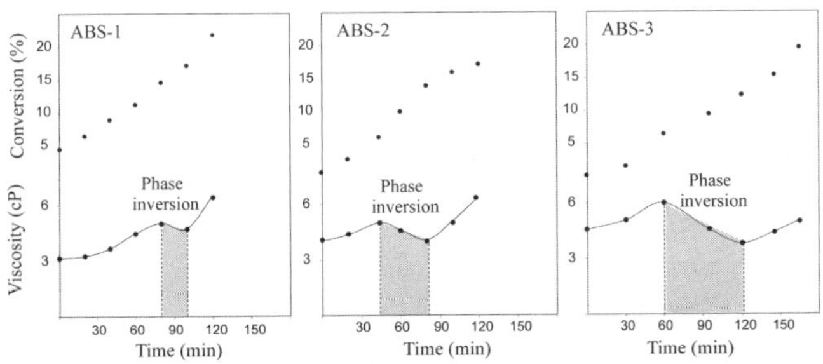

Figure 2. Evolution in the conversion and viscosity as a function of polymerization time for the three synthetized ABS polymers.

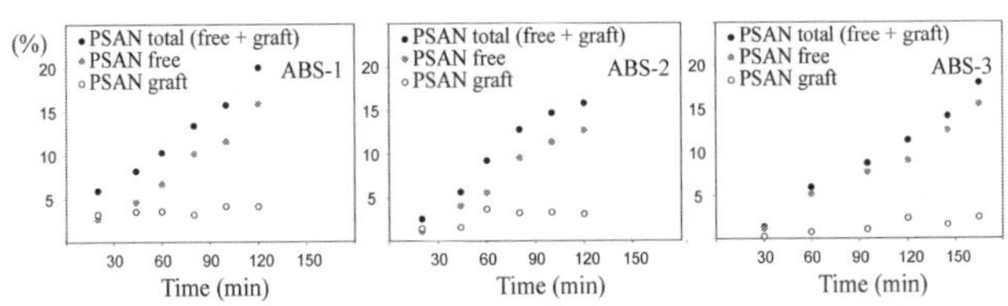

Figure 3. Behavior of the produced SAN as a function of the polymerization time.

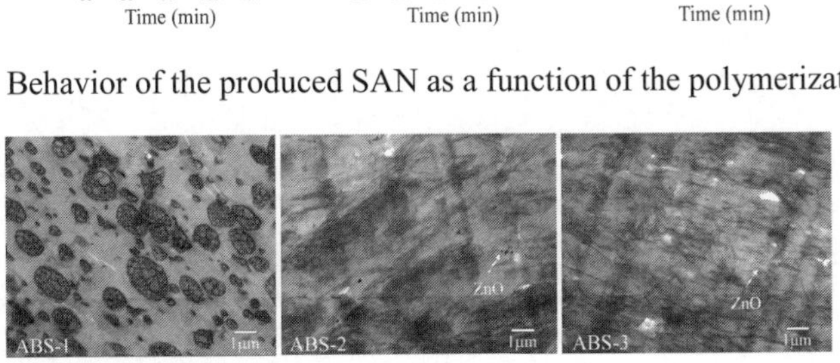

Figure 4. Final morphology of the three synthetized ABS´s.

Table 2. Characteristics and properties of the different ABS.

Material	IS (ft-lb_f/in)	E (MPa)	σy (MPa)	X (%)	Mw (kg/mol)
ABS-1	4.00	745.7	24.04	14.56	308
ABS-2	6.64	856.7	25.83	13.90	349
ABS-3	5.15	730.4	28.98	15.34	408

IS: Impact strength; E: Young's modulus; σy: yield strength; X: conversion; Mw: weight average molecular weight.

Conclusions

The incorporation of nano-ZnO during the ABS synthesis modified the final ABS morphology transforming the conventional salami morphology highly occluded with SAN to a highly dispersed morphology in the SAN matrix. The total production of SAN (free and grafted) was adversely affected by the presence of the nanoZnO. The modified nano-ZnO with an aminosilane compound had the major influence on the total SAN production. Our work, also confirmed that the nano-ZnO improved the mechanical properties of ABS.

Acknowledgements

The authors thank CONACYT for financial support, of projects CB-007/80159 and CB-007/80158.

References

[1] Haaf F., Breuer H., Achte A., Schmitt B.J. and Stabenow J., *J. Sci. Ind. Res*, **40**, 659-679 (1981).

[2] Maul J., *Ullmanns's Encyclopedia of Industrial Chemistry*. 6th Ed.; Wiley – VCH, 2002

[3] Chul Kim B. and Jae Lee S., *Korea-Australia Rheol. J.*, **20**, 227-233 (2008)

[4] Ramón Tena Zaera, " Zinc Oxide: Crystalline growth by gas phase transportation and characterization of physical properties ", Ph. D. Dissertation (2004). Spanish Language.

[5] Luna Comparán Nidia "Synthesis and characterization of ZnO nanoparticles in Polystyrene Matrix by emulsion polymerization technique" Bachelor thesis. (2009). Spanish Language.

Keyword Index

Author Index